Neuroc
Volume

Oxford Series In Developmental Cognitive Neuroscience

Series editor: Mark H. Johnson, *Centre for Brain and Cognitive Development, Birkbeck College, University of London, UK*

Books in this series:

Neuroconstructivism: Volume One—How the Brain Constructs Cognition

Denis Mareschal, Mark H. Johnson, Sylvain Sirois, Michael W. Spratling, Michael S.C. Thomas, and Gert Westermann

Neuroconstructivism: Volume Two—Perspectives and Prospects

Edited by Denis Mareschal, Sylvain Sirois, Gert Westermann, and Mark H. Johnson

Neuroconstructivism Volume Two

Perspectives and Prospects

Edited by

Denis Mareschal
Centre for Brain and Cognitive Development,
Birkbeck College, University of London, UK

Sylvain Sirois
School of Psychological Sciences,
University of Manchester, UK

Gert Westermann
Department of Psychology,
Oxford Brookes University, Oxford, UK

Mark H. Johnson
Centre for Brain and Cognitive Development,
Birkbeck College, University of London, UK

OXFORD
UNIVERSITY PRESS

OXFORD
UNIVERSITY PRESS

Great Clarendon Street, Oxford OX2 6DP

Oxford University Press is a department of the University of Oxford.
It furthers the University's objective of excellence in research, scholarship,
and education by publishing worldwide in

Oxford New York

Auckland Cape Town Dar es Salaam Hong Kong Karachi
Kuala Lumpur Madrid Melbourne Mexico City Nairobi
New Delhi Shanghai Taipei Toronto

With offices in

Argentina Austria Brazil Chile Czech Republic France Greece
Guatemala Hungary Italy Japan Poland Portugal Singapore
South Korea Switzerland Thailand Turkey Ukraine Vietnam

Oxford is a registered trade mark of Oxford University Press
in the UK and in certain other countries

Published in the United States
by Oxford University Press Inc., New York

First published 2007

British Library Cataloguing in Publication Data
Data available

Library of Congress Cataloging in Publication Data
Data available

Typeset by SPI Publisher Services, Pondicherry, India
Printed in Great Britain
on acid-free paper by
Biddles Ltd., King's Lynn, Norfolk

ISBN 978–0–19–852991–0 (Volume 1 Paperback)
ISBN 978–0–19–852990–3 (Volume 1 Hardback)
ISBN 978–0–19–852993–4 (Volume 2 Paperback)
ISBN 978–0–19–852992–7 (Volume 2 Hardback)
ISBN 978–0–19–921482–2 (Volumes 1 and 2 Paperback Set)

1 3 5 7 9 10 8 6 4 2

To those we love—families and friends—who have continually supported us and enabled us to follow our dreams.

Acknowledgements

The work we describe in these two volumes and the writing of the volumes themselves could not have occurred without the material support of a number of generous benefactors. First we wish to thank Birkbeck and the MRC for their generous support of the Centre for Brain and Cognitive Development. The book has also benefited directly from awards made by the European Commission (HPRN-CT-2000-00065; 516542-NEST) and the Medical Research Council UK (G9715887). Finally, SS was supported by a Fellowship from NSERC Canada. We also thank Tom Shultz for graciously reading through portions of this book.

List of Contributors

James A. Bednar
Institute for Adaptive and Neural Computation
University of Edinburgh

Ira L. Cohen, Ph.D.
Chairman, Department of Psychology
NYS Institute for Basic Research in Developmental Disabilities

Yiannis Demiris
Department of Electrical and Electronic Engineering
Imperial College London

Melissa Domingez
DBK Acoustics

Robert A. Jacobs
Department of Brain and Cognitive Sciences
University of Rochester

Marc Joanisse
Department of Psychology and Program in Neuroscience
The University of Western Ontario

Mark H. Johnson
Centre for Brain and Cognitive Development
Birkbeck College,
University of London

Denis Mareschal
Centre for Brain and Cognitive Development
Birkbeck College,
University of London

Risto Miikkulainen
Department of Computer Sciences
The University of Texas at Austin

Shreesh P. Mysore
Control and Dynamical Systems Program
California Institute of Technology

Domenico Parisi
Institute of Cognitive Sciences and Technologies
Italian National Research Council

Steven R. Quartz
Division of the Humanities and Social Sciences, and
Computation and Neural System Program
California Institute of Technology

Maartje Raijmakers
Department of Psychology
University of Amsterdam

Matthew Schlesinger
Brain and Cognitive Sciences Program
Department of Psychology
Southern Illinois University Carbondale

Thomas R. Shultz
Department of Psychology and School of Computer Science
McGill University

Sylvain Sirois
School of Psychological Sciences
The University of Manchester

Olaf Sporns
Department of Psychological and Brain Sciences
Indiana University Bloomington

Gert Westermann
Department of Psychology
Oxford Brookes University

Contents

Chapter 1

Introduction

Denis Mareschal, Sylvain Sirois, and Gert Westermann

All science is either physics or stamp collecting.

Ernest Rutherford (1871–1937) Nobel Prize-winning chemist and physicist, cited in Birks (1962)

The real challenge for developmental psychology is to understand the *how* of development, and not just the *what* of development. That is, we need to have causal theories of what makes the behaviours that are of psychological relevance emerge. These theories need to explain behaviours at multiple timescales. They must explain how and why behaviours unfold online, as we observe them in real time, as well as how they unfold in developmental time—across the lifespan of the individual.

To do this, we need more than just a very detailed description of the behaviours that can be observed at any point in time. Certainly, such descriptions are essential for the advancement of causal theories of development, but they are unsatisfactory to the extent that they are unable to explain or predict new behaviours. The clearest example of this is with regards to explaining the behaviour of children with developmental disorders. Unless one has a mechanistic theory of what is causing behaviours to unfold, and a causal theory of what is atypical in such children's processing, it is impossible to explain or predict why one set of atypically developing children may show a delay at one behaviour and excel at another, while a second set of atypically developing children will excel at the former behaviour but have a developmental delay in the latter behaviour.

Critics of this view could argue that as we discover new and unexpected behaviours in different populations or in different contexts, we can always update our detailed descriptions. However by going down such a route, we come dangerously close to what Rutherford called 'stamp collecting'. Lord Rutherford was a physical chemist at a time when physics was making great strides in uncovering the mysteries of how the physical world works. His assertion cited at the beginning of this chapter is more than just chauvinism. It reflects the deep belief that without a causal theory of what links the observables around us, a theory that is grounded in mechanisms and processes whose effects can be measured, then we are not doing proper science. This volume contains ten concrete examples of how such causal theories can be developed in the study of neuroconstructivist cognitive development.

In Volume 1, *Neuroconstructivism: How the Brain Constructs Cognition*, we show how neural representations that support cognition emerge and are sustained in the brain. The key concept on which the emergence of representations is predicated is that of *context-dependence*. Whether it be at the level of individual neurons acquiring their eventual functions during early development, at the level of functional brain systems, at the level of the whole brain embedded within a physical body, or whether it was at the level of the individual child situated within a physical and social environment, *development does not occur in a vacuum*. The context within which an entity exists, and its interactions with its neighbours, are fundamental in determining the kinds of functions that it will take on and the kinds of representations that will develop. This is true at all levels of investigation. General processing constraints, operating at different levels of description, shape the emergence of what we have called *partial representations*, in which only minimal representations that complement information in other functional brain systems or in the environment are required and therefore acquired. As such, representations are fundamentally context-dependent. The notion of partial representations can be contrasted to the idea that the brain contains fully detailed singular representations of the environment.

These constraints can be observed operating at the level of neural information processing. Ultimately, the brain is made up of neurons. The way information is processed in complex neural systems dictates what kind of information these representations can support, and the way that the representations will interact causally with each other. The constraints are also observed when considering that the brain is embodied within a physical body that is itself situated within a physical and social environment. The body and the environment together constrain the types of problems that the child will be confronted with but they are also potential sources of solutions to those same problems. That is, the brain–body–environment complex can make use of the body and the environment as integral components of the information processing solutions to particular problems. Similarly, brain regions or functional systems do not exist in a vacuum. Instead, they exist within the context of other brain regions or functional brain system: thus functional brain systems are both constrained by and can capitalize on neighbouring functional brain systems. We have called this notion 'embrainment'.

More specifically, we have identified three general mechanisms that operate in the emergence of representations: *competition*, *cooperation* and *chronotopy*. Representations (and in turn the behaviour that depends on these representations) emerge from a self-organizing process in which many components compete and cooperate. Thus, multiple components contribute to any specific representation. There can also be multiple different instantiations of a representation and these can differ dynamically depending on the context in which the representation emerges. Finally, because development is integrative and builds on previous representations, representations become increasingly fragmentary and increasingly linked to other representations with age of acquisition.

We have also identified two general processes involved in the emergence of representations: *proactivity* and *progressive specialization*. Representations arise as a result of internally generated activity. This is true at the level of individual neurons, cell assemblies, functional brain units, or whole organisms. The child actively queries the environment by generating behaviours that result in particular consequences and, does not just passively internalize an external world. The emergence of representations is strongly constrained by the ontological history of the individual. One way in which this can be realized is through the operation of chronotopic constraints. This is the idea that adaptive plasticity occurs at different times in different regions of the neural system. Progressive specialization has two consequences: (1) the fact that what happens now is limited by what has happened in the past, and (2) the fact that early representations also serve as building blocks for later representations.

Because development is a very complex, context-dependent process that unfolds through time, it is unlikely to be understood simply by analysis or experimental observation. While experimental observation has generated a great deal of data, our ability to understand and relate these otherwise isolated facts is severely limited. Mechanistic theories are therefore required to help us reason about such complex phenomena, to interpret, and to make sense of experimental data.

Mechanistic or process theories can often be translated into models. A model emulates, mimics, depicts or exemplifies (certain aspects of) something else—a computational model does so by simulation. An implemented computational model has two clear advantages over a theory (Lewandowsky, 1993).

First, instantiating theories in sufficient mathematical detail that they can be implemented in the form of a computational simulation requires a level of specification that typically eludes verbal theories. A model can thus provide a precise formulation of an otherwise underspecified theory. Furthermore, since computational models are built on mathematical laws, the principles of their operation are explicit. This rigorous declaration of a theory can enable more accurate communication of ideas and reduce the scope for misinterpretation. In addition, the process of translating a theory into a model can, in itself, be beneficial since this exercise can uncover previously ignored gaps or inconsistencies in the theory.

A second advantage of modelling is that the model itself can become the subject of investigation. A computational model instantiates a formal (usually mathematical) description that allows the biological system to be simulated. Such simulation not only enables theories to be checked and tested, but can provide a tool for exploring the theory and making comparisons between competing theories. A simulation can be used to test a theory in ways that are otherwise beyond the scope of experimental investigation, or to explore systems that would otherwise be too complex. Crucially, working models can generate predictions that can guide future experimental work. This process of modelling, generating predictions and testing these predictions, can lead to improvements in the model as well as in the empirical database and hence help to synthesize more complete theories.

In order for a model to have explanatory power it needs to be specified at an appropriate level of abstraction. What constitutes the correct level of abstraction will depend on the phenomenon under investigation, the aims of the research, and the audience at whom it is aimed. A model at too high a level of abstraction might provide a good description of the behaviour of the system being modelled but fail to provide any information about the mechanisms by which that system operates. A more detailed, functional model, which reveals the principles of operation of the system, and which can thus explain the observed behaviour, should be preferred over a purely descriptive model. However, a model at too low a level of abstraction may be so complex that it fails to identify those principles that are crucial to the modelled systems' behaviour or function. A model needs to provide some simplification of reality, otherwise little is gained in creating it and the general, explanatory, principles will remain hidden behind extraneous detail. There is thus a trade-off between simplifying reality (to obtain an informative and practical model) and faithfulness to reality (to obtain an accurate and complete model). A useful model must strive to include all the relevant principles, while leaving out all the irrelevant ones (e.g., Rumelhart and McClelland, 1986).

Computational- and numerical-modelling are well-established tools in the physical and neural sciences. We strongly believe that psychology is a mature enough science (in terms of having a sufficiently large empirical database) for modelling to make an equivalently important contribution. Indeed, this is a belief shared by many other authors (e.g., Elman *et al.*, 1996; Klahr and Wallace, 1978; Shultz, 2003; Simon, 1962).

In Volume 1, we alluded to the importance of models in several places, and even sketched a few key models that illustrated the processes that we were discussing. In this volume, nine different research labs with objectives broadly consistent with the neuroconstructivist approach present their computational modelling work. Although the chapter authors may not agree with all the views expressed in Volume 1, we invited them to contribute to this volume because we feel that their work constitutes an explicit example of how neurocomputational methods can be used to investigate the causal mechanisms of adaptation (learning and development). The questions that the models are built to investigate differ both in the level of description and in the timescale over which the relevant behaviours operate. Some models focus on relatively rapid adaptation occurring (perhaps) at the cellular level of descriptions, whereas other models focus on relatively slow adaptation occurring (perhaps) at the cognitive level of adaptation.

We have asked all contributors to emphasize the following aspects of their contributions when describing their work:

1. What functional brain constraints operate on the process of representation development?
2. What embodiment or situatedness constraints operate on the process of representation development?

Indeed, one of the key conclusions of Volume 1 is that the representations that underlie cognition are constrained both by the substrate in which they are implemented (i. e., the neural networks of the brain) and the physical reality of the body and environment in which the brain is situated. Thus, we argued for the rejection of the Marrian idea that levels of description can be studied independently. We believe that the models described in the chapters to follow are consistent with this vision.

Ultimately, computational models are tools to help us reflect on questions of process and mechanisms. Therefore, we have also asked the authors to answer the following two questions:

3. How does the model embody these constraints?
4. What concrete predictions does the model make?

These are important questions because it is vital that readers understand why a model works. If a model remains as opaque as the phenomenon it is meant to clarify then it has little value. Moreover, all models should be testable. To be falsifiable it must relate to data in such a away that its predictions can be tested.

The first chapter, written by Bednar and Miikkulainen, examines the extent to which visual representations can begin to emerge prenatally. This is a fascinating question because while the unborn child exists in a relatively rich auditory environment, its visual environment is very limited, consisting essentially of very coarse variation in luminosity levels. However, there is evidence of substantial spontaneous intrinsically generated neural activity in the visual cortex and elsewhere in the brain both pre- and postnatally. This intrinsic activity can begin to shape the way in which visual information will be represented once the infant is born and experiences the visual world directly. These authors describe a neural network model that can explain the very rapid emergence of orientation feature detectors and face sensitive detectors after only very limited direct visual experience as a result of the intrinsic activity 'setting the stage' for the kinds of neural representations that are allowable and direct visual experience. The model illustrates the difficulties, from a developmental perspective, of differentiating too clearly between pre- and postnatal development. Both pre- and postnatal exogenous events combine to explain the emergence of the representation that we then observe after birth.

The subsequent four chapters can be construed as addressing a common theme. All four ask how 'development' can interact with 'learning' to produce the pattern of behavioural change that we observe in the growing child. As discussed in Volume 1, dichotomizing learning and development as different kinds of things is highly controversial. There is little agreement over what might count as an example of development as opposed to learning and vice versa (Liben, 1987; Lewkowicz and Licklister, 2002). One way to try to address this question is to implement in a computational model the different mechanisms that might underlie these different kinds of events and explore how those hypothesized mechanisms would interact in real time.

In Chapter 3, Domingez and Jacobs ask how visual development might interact with visual learning to explain patterns of motion detection and stereoscopic vision that we see unfolding in the developing infant. Their model examines more specifically the 'less-is-more' hypothesis in the domain of visual development. This is the idea that initially limiting the amount of information that the developing infant has access to (perhaps through limited visual acuity) actually improves long-term learning because it allows the infant to extract the central tendencies of a domain before learning the fine detail (Turkewitz and Kenny, 1982). This is an illustration of the chronotopic processes discussed in Volume 1.

To a degree, this model focuses more on the mechanisms of learning and does not directly address what causes development: development is simply implemented by gradually increasing the spatial frequency resolution of the events from which the model learns stereopsis and motion perception. This is analogous to increasing the visual acuity of the infant. In the human infant, this is known to happen largely because of changes in the shape of the eye and the number and location of photoreceptors in the retina in the first few months of life (Banks and Shannon, 1993). This chapter also illustrates one of the strengths of the modelling approach: models can be used to explore hypothetical scenarios that are not found in the real world. Here the authors test the 'more-is-less' hypothesis by evaluating the performance of models under different developmental schedules. They compare models with different schedules of access to the full range of spatial frequencies (e.g., high-to-low transitions vs. low-to-high transitions). In each case, they find that a gradual increase in the range of spatial frequencies accessible to the visual system (from low first to high later) leads to the best eventual performance of the model.

In Chapter 4 Shultz, Mysore and Quartz ask more directly how development can be conceptualized within the context of neural networks that learn. They argue that experience-dependent architectural plasticity (i.e. experience-dependent growth of the neural tissue through processes such as synaptogenesis) has been largely under-emphasized in current accounts of neural network learning. They use the example of auditory localization in barn owls as a case study in which architectural adaptation plays a fundamental role. The presence of architectural plasticity also has consequences for models of higher-level cognitive abilities. Networks that are allowed to grow as part of their adaptation can gradually increase in computational power (i.e., in the kinds of things that they can come to represent). Thus, they provide a natural tool for modelling cognitive development across longer age ranges. Shultz *et al.* show how the casacade correlation learning architecture—a neural network architecture that can adjust both its weights and architecture as part of learning—(Fahlman and Lebierre, 1990) can be used to model a broad range of developmental phenomena in children's reasoning. In particular, the authors argue that architectural plasticity may underlie what Piaget described as stages of development—the observation that through childhood there are periods with little apparent development alternating with periods of very rapid development.

Raijmakers adopts a different solution to the question of stage transitions in development in Chapter 5. Her contribution is unique in emphasizing the dynamic nature of the neural information processing that underlies cognitive development. She makes the point that what might appear from the outside as a sudden transition in performance, and which may therefore be interpreted as evidence of a radical restructuring of the representations that underlie cognition, can arise from the continuous adjustment of the underlying control parameters of a system. This idea comes out of the physical sciences where there are many examples of systems in which a gradually changing parameter leads to a rapid, apparently qualitative, change in the state of the system (for example, the gradual increase in the temperature of water will lead it to transition from solid to liquid at 0° Celsius and from liquid to gas at 100° Celsius). Raijmakers argues that this is, in fact, what is happening when we observe stage transitions during children's increases in performance on specific tasks. New evidence examining the details of children's responses suggests that many of them show 'catastrophe flags' characteristic of such transitions at critical developmental junctions. She goes on to show that these kinds of transitions can take place in recurrent neural networks in which activation is allowed to spread in all directions. This is demonstrated using an Adaptive Resonance Theory (ART) network (Grossberg, 1980) network to model discontinuities in concept learning.

In Chapter 6, Schlesigner and Parisi emphasize that adaptation can occur at a developmental timescale—within the lifetime of an individual—but also across individuals (phylogeny). To explore these interactions they use a modelling method referred to as 'genetic algorithms' (Mitchell, 1996). This method combines the simulation of learning within the individual using standard neural network technique with an evolutionary selection process in which successive 'generations' of networks gradually evolve to change the learning properties of the individuals as a result of longer-term adaptive pressures. An important point made in this chapter is that while this kind of mechanism can be used to model processes occurring across the lifetimes of many individuals, it can also be used to model the slow evolution of neural or cognitive structures within the timescale of an individual developing child. Schlesinger and Parisi illustrate this point with models of the development of reaching, the acquisition of object permanence in infants, and aspects of language development.

The next two chapters also share a common theme in that they emphasize the need to take constraints from embodiment seriously. Here, instead of simply simulating interactions within a computer program, the computer programs are used to control a real robot that interacts with the world in real time. The solutions that the model acquires to solve certain key problems are constrained by the structure of the world that it is navigating through and the physical properties of its body.

Chapter 7, by Demiris, describes his work on explaining the emergence of imitation in infants using neurally plausible mechanisms embodied within a series of progressively more complex robots that learn to imitate. The model is designed to implement and test Meltzoff and Moore's (1997) *active intermodal matching* account of imitations. In a first example Demiris illustrates how knowledge about the spatial organization of the environment in which the robots navigate can be communicated between robots using imitation. The second example illustrates how a particular developmental theory (the Intermodal Matching Hypothesis) can be made more specific by implementing it in a functional computational system. Finally, the third example discusses how physiological data on the functional properties of mirror neurons can be used to constrain the kinds of mechanisms that operate within a robot learning imitation.

The importance of biologically plausible mechanisms in developing robot models is emphasized even more by Sporns in Chapter 8. He also argues for the importance of robotic models that are controlled by artificial neural networks. Such models provide a tangible way of investigating how the body, the environment and the neurocontrol system of the brain co-determine a child's unfolding abilities. He illustrates this with a robotic model that learns to navigate and interact with objects in the world. Building such systems underscores the importance of attributing differential 'value' to different objects or actions on the world for obtaining a realistic developmental profile. Sporns hypothesizes that in terms of neural information processing 'value' can be linked to differing levels of dopamine in the brain associated with particular events and tasks. Dopamine is a neuromodulator that effects the level of plasticity in a neural network.

Finally, the last two chapters illustrate how computational models can be used to understand what causes the emergence of atypical behaviours observed in children with developmental disorders. In Volume 1, we argue that developmental disorders can be viewed as the outcomes of typical developmental processes unfolding under atypical constraints. Both of these chapters examine how very specific modifications to the input or to one of the processing constraints of neural networks can give rise to the selective patterns of behaviour observed in children.

In Chapter 9, Joanisse examines what might be some of the causes of the language impairments observed in children with specific language impairments (SLI). These are children who appear to function in the normal range on a broad range of cognitive tasks, but who have very poor language skills. One argument has been that such children have a selectively impaired language module. However, an alternative hypothesis is that such children have subtle impairments in the detection or processing of very low-level auditory features (e.g., rapidly varying temporal variations of auditory information). As a result, they have poorer phonological discrimination abilities which in turn effects their ability to process subtle syntactic markers associated with a specific morpheme (e.g., the –ed added to the end of regular verbs to form the past tense in English). Joanisse argues that these low-level deficits, coupled with the fact that language acquisition is a cumulative developmental process (in which early deficits can have knock-on effects on later development), are the real causes of these language impairments. Neural network models can be used to illustrate how early low-level auditory processing deficits can lead to specific syntactic deficits in later development. From the neuroconstructivist perspective, the important conclusion here is that high-level cognitive abilities (syntax) are constrained by the low-level input constraints of the body (e.g., auditory discrimination skills). This illustrates why understanding how the mind is embodied is essential for understanding the causes of the behaviours that we are studying.

Chapter 10 emphasizes how models can be used to make predictions about how children with developmental disorders will behave. Cohen describes a simple model of the learning styles and savant skills of children with autism and argues that the different learning styles reflect an increased number of neurons in functional brain systems in autism. While this model does a good job of capturing the broad facets of learning in children with autism, it also makes specific predictions about behaviours in novel contexts and about the kinds of interventions that are likely to be most effective given the neural mechanisms that underlie the learning styles observed. Thus, this chapter illustrates how one can make the transition from theory to practice using implemented models.

Before turning to the chapters themselves, we wish to make one or two procedural points that should help with the reading of this book. The chapters have been written to be accessible and of interest to a very broad range of readers, some of whom may be very technically sophisticated, whereas others may be somewhat less experienced with the terminology of computational modelling. As a result, we have asked the authors to place the more technical details of the models they described inside boxes within each chapter. Those who wish to know the technical details can choose to read the boxes to evaluate the models more closely; however, the chapters are meant to be readable without reference to the technical detail. Readers will also notice that the level of technical sophistication of the models described differs from chapter to chapter and from domain to domain. This reflects the history of using models as a research tool in each of these subdomains. Whereas the visual sciences have been using models for many years now (and therefore are used to producing more technically dense models), only relatively recently have computational models been used to study developmental disorders. Finally, because of the broad range of domains covered, each chapter contains a small glossary at the end to help clarify terms that may seem perfectly clear to the initiated, but obscure and jargon-like to the novice. Our ultimate aim in brining together these chapters is to provide an enjoyable and informative introduction to modelling that illustrates the principle of the neuroconscructivist approach to development.

References

Banks, MS and Shannon, E (1993). Spatial and chromatic visual efficiency in human neonates. In CE Granrud (ed.) *Visual perception and cognition in infancy*, pp. 1–46, Hillsdale, NJ: Erlbaum.

Birks, JB (1962). Rutherford at Manchester. London, UK: Heywood.

Elman, J, Bates, E, Johnson, MH, Karmiloff-Smith, A, Parisi, D and Plunkett, K (1996). *Rethinking innateness: A connectionist perspective on development.* Cambridge, MA: MIT Press.

Fahlman, SE and Lebiere, C (1990). The cascade-correlation learning architecture. In DS Touretzky (ed.) *Advances in neural information processing systems* 2, pp. 524–532. Los Altos, CA: Morgan-Kaufmann.

Grossberg, S (1980). How does a brain build a cognitive code? *Psychological Review*, 87, 1–51.

Klahr, D and Wallace, JG (1978). *Cognitive development: An information processing view.* Hillsdale, NJ: Erlbaum.

Lewandowsky, S (1993). The rewards and hazards of computer simulations. *Psychological Science*, 4, 236–243.

Lewkowicz, DJ and Licklister, R (2002). *Conceptions of development.* London: Psychology Press.

Liben, LS (1987) *Development and learning: Conflict or congruence.* Hillsdale, NJ: Erlbaum.

Meltzoff, AN and Moore, MK (1997). Explaining facial imitation: A theoretical model. *Early Development and Parenting*, 6, 179–192.

Mitchell, M (1996). *An introduction to genetic algorithms. Cambridge*, MA: MIT Press.

Rumelhart, DE and McClelland, JL (eds) (1986). *Parallel distributed processing: Explorations in the microstructure of cognition. Volume 1: Foundations.* Cambridge, MA: MIT Press.

Shultz, TR (2003). *Computational developmental psychology.* Cambridge, MA: MIT Press.

Simon, H (1962). *An information processing theory of intellectual development.* Monographs of the Society for Research in Child Development. Anne Arbor, Michigan: SRCD.

Turkewitz, G and Kenny, PA (1982). Limitations on input as a basis for neural organization and perceptual development: a preliminary theoretical statement. *Developmental Psychobiology*, 15, 357–368.

Chapter 2

Constructing visual function through prenatal and postnatal learning

James A. Bednar and Risto Miikkulainen

Introduction

Specific regions in the visual cortex of adult humans and other adult mammals selectively respond to different visual features. For instance, the primary visual cortex (V1) is arranged into an orientation map, consisting of an array of neurons that respond to visual contours with certain orientations, but not to others (Figure 2.1) (Blasdel, 1992; Hubel and Wiesel, 1962, 1968). Similarly, visual areas in the fusiform gyrus respond more strongly to human faces than to other complex stimuli (Kanwisher, McDermott and Chun, 1997). How these selective responses develop is not yet clear. On the one hand, experimental and computational results (reviewed later) suggest that much of the structure and function of the visual system is constructed by a general-purpose learning process, driven by inputs from the environment. On the other hand, there is also considerable evidence that many aspects of the visual system are hard-wired, i.e., constructed from a specific blueprint encoded in the genome.

The conflict between these two positions is generally known as the Nature–Nurture debate, which has been raging for centuries in various forms (Diamond, 1974). The idea of a specific blueprint does seem to apply to the largest scale organization of the visual system, at the level of visual areas and their interconnections. These patterns are largely similar across individuals of the same species, and their development does not generally depend on neural activity, visually evoked or otherwise (Miyashita-Lin, Hevner, Wassarman, Martinez, and Rubenstein, 1999; Rakic, 1988; Shatz, 1996). But at smaller scales, i.e., neurons and connections within visual areas, there is considerable evidence for both environmental and internally controlled development. Thus, debates centre on how this seemingly conflicting evidence can be reconciled.

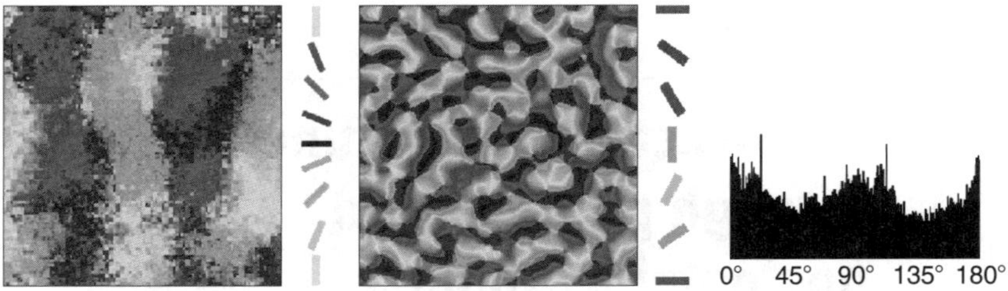

Fig. 2.1 Newborn and adult animal orientation maps. Plot (a) shows a 1.9 × 1.9 mm area of newborn cat V1, and (b) shows a 5 × 5 mm area of adult monkey V1, both measured by optical imaging techniques. (a) Reprinted with permission from Crair *et al.* (1998), *Science*, 279, 566–570, copyright 1998 American Association for the Advancement of Science; (b) reprinted with permission from Blasdel (1992), *Journal of Neuroscience*, 12, 3139–3161, copyright 1992 by the Society for Neuroscience). Each neuron in the maps is coloured in grey scale according to the orientation it prefers, using the key adjacent to it. Nearby neurons in the map generally prefer similar orientations, forming groups of the same colour. Other qualitative features are also found, such as pinwheels, i.e. points around which orientation preference changes continuously. Newborn maps have similar qualitative features, although they are noisier. (c) The distribution of orientation preferences has been measured in ferrets, and has a bias towards vertical (90) and horizontal (0 and 180) contours, which reflects the statistics of the environment. (c reprinted with permission from Coppola *et al.*, 1998, PNAS 95, 2621–2623, copyright 1998 National Academy of Sciences, USA.).

In this chapter, we will briefly give an overview of the relevant areas of the visual system, and summarize the evidence for environmental and genetic influences on orientation and face processing. We will then propose a simple hypothesis about how the cortex can combine both influences, through prenatal learning of spontaneous activity patterns followed by postnatal learning of visual stimuli. Using a computational model of orientation and face processing, we will show that this hypothesis can explain a large body of data on how low-level (i.e. orientation) and higher-level (i.e. face) perception develop. The orientation map simulations are strongly constrained by existing experimental data, providing a rigorous way to test and calibrate the model, while the face-processing simulations will provide a set of concrete predictions for future experiments.

Biological and psychological background

The human visual system, like that of most mammals, consists of a series of topographically organized visual areas (for review, see e.g. Kandel, Schwartz and Jessell, 1991). When light falls on the retina, photoreceptor cells convert the light patterns into patterns of electrical activity. Neurons in the retina called retinal ganglion cells have receptive fields (RFs) covering a local patch of photoreceptors, and are active when the light patterns are a good match to their RFs. The retinal ganglion RFs tend to have a centre–surround organization, responding to light dots surrounded by dark (ON retinal ganglion cells), or dark patches surrounded by light (OFF retinal ganglion cells). Retinal ganglion cell activity is relayed through the optic nerve to ON and OFF neurons in the LGN (lateral geniculate nucleus of the thalamus), which have RFs very similar to retinal ganglion cells. The LGN cells relay the activity patterns to the primary visual cortex (V1), which in the adult has RFs selective for oriented patterns, instead of the circular RFs typical of earlier regions. As mentioned earlier, these RFs form a topographically organized orientation map, representing the orientation of each small patch of the input image. The activity levels of neurons in the orientation map are relayed to neurons elsewhere in the visual cortex, which have RFs selective for more complex objects such as human faces.

The development of the eye, LGN, and V1 have been studied extensively, in part because these areas are very similar across many species. However, relatively little is known about how higher cortical areas develop. As described in the next subsection, experiments with V1 suggest that orientation maps develop from both environmental (i.e., external) and internal cues. Face processing is much more difficult to study, requiring higher primate or human subjects. Thus most of the available face-processing data is from behavioural studies of human infants, rather than detailed measurements of individual neurons and neural regions. These data suggest that face processing may develop similarly to orientation processing, driven by both environmental and internal sources (as described in the second subsection). A third subsection describes how internally generated activity could contribute to the development of these capabilities, based on recent experimental discoveries of widespread spontaneous activity.

Orientation map development

Experiments in monkeys, ferrets and cats suggest that orientation-selective responses in adults could be learned from exposure to oriented patterns (as reviewed by Movshon and van Sluyters, 1981). For instance, Blakemore and Cooper (1970) found that if kittens are raised in environments consisting of only vertical contours, most of their V1 neurons become responsive to vertical orientations. Similarly, orientation maps from kittens with such rearing devote a larger area to the orientation that was overrepresented during development (Sengpiel, Stawinski, and Bonhoeffer, 1999). Conversely, kittens raised without patterned visual experience at all, e.g. by suturing their eyelids shut, have few orientation-selective neurons in V1 as an adult (Blakemore and van Sluyters, 1975; Crair, Gillespie, and Stryker, 1998). Even in normal adult animals, the distribution of orientation preferences is slightly biased towards horizontal and vertical contours (Chapman and Bonhoeffer, 1998; Coppola, White, Fitzpatrick, and Purves, 1998). Such a bias would be expected if the neurons learned orientation selectivity from typical environments, which have more horizontal and vertical contours than oblique ones, for example trees and horizons (Switkes, Mayer, and Sloan, 1978).

In ferrets, it is even possible to reroute the connections from the eye that normally go to V1, so that instead they reach the auditory cortex (as reviewed in Sur, Angelucci, and Sharma, 1999; Sur and Leamey, 2001). As a result, the auditory cortex develops orientation-selective neurons and orientation maps, although these structures are not as pronounced as in normal maps.

Furthermore, the ferret can use the rewired neurons to make visual distinctions, such as discriminating between two grating stimuli (von Melchner, Pallas, and Sur, 2000). These experiments show that the input to a cortical area can profoundly affect its structure and function. However, there is also significant evidence suggesting that V1's structure is genetically determined. For example, it has long been known that individual orientation-selective cells exist in newborn kittens and ferrets even before they open their eyes (Blakemore and van Sluyters, 1975; Chapman and Stryker, 1993). Psychological studies further suggest that human newborns can already discriminate between patterns based on orientation (Slater and Johnson, 1998; Slater, Morison, and Somers, 1988). Recent advances in experimental imaging technologies have made it possible to measure the full map of orientation preferences in young animals. Such experiments show that large-scale orientation maps exist prior to visual experience, and that these maps have many of the same features as found in adults (Chapman, Stryker, and Bonhoeffer, 1996; Crair *et al.*, 1998; Gödecke, Kim, Bonhoeffer, and Singer, 1997). Furthermore, the global patterns of orientation-selective patches in the maps appear to change very little with normal visual experience, even as the individual neurons gradually become more selective for orientation (Chapman and Stryker, 1993; Crair *et al.*, 1998; Gödecke *et al.*, 1997). Thus, despite the clear influence of environmental input on visual cortex structure, normal visual experience appears primarily to preserve and fine-tune the existing structures, rather than drive their development.

Taken together, the evidence indicates that both genetic and environmental influences interact to produce the adult orientation map. However, important questions remain. How does this interaction actually occur? Could adult-like maps develop from environmental or genetic cues alone, or are both necessary?

These questions are difficult to answer through biological experiments. Computational modelling, however, can lead to valuable insights, because it is easy to separate environmental and genetic influences in computational experiments. Existing models have been used to simulate how orientation maps can develop from visual input alone, (based on natural images; Burger and Lang, 1999) or genetic factors alone, (based on noise representing spontaneous neural activity; Linsker, 1986: for review of existing models of each type, see Swindale, 1996.) However, prior models have not shown how V1 can have an initial map at birth that becomes smoother and more selective due to postnatal visual experience, while retaining the original map shape.

Previous models driven by visual or internal activity alone also differ in many ways besides the source of activity, and thus it has been difficult to determine whether the activity patterns alone account for any differences between the results. This chapter will present results that show clearly how orientation maps can develop both before and after birth in the same model.

Development of face detection

Face perception appears to develop similarly to orientation perception, with evidence for both postnatal learning and function at birth. A number of studies have found that newborns prefer face-like patterns (Goren, Sarty, and Wu, 1975; Johnson, Dziurawiec, Ellis, and Morton, 1991; Simion, Valenza, and Umiltà, 1998), and others show that newborns can discriminate between specific faces and learn to prefer a particular face within the first few hours and days after birth (Bushnell, 2001; Pascalis, de Schonen, Morton, Deruelle, and Fabre-Grenet, 1995). When shown moving schematic faces in the visual periphery, newborns and one-month-olds will follow them further than other similar patterns (Goren *et al.*, 1975; Johnson *et al.*, 1991) (see example schematics in Figure 2.6 b–e). However, these abilities change significantly over early development. Older infants do not significantly prefer schematic faces in the visual periphery (Johnson *et al.*, 1991), but between one and two months they begin to respond to facial features in central vision (Maurer and Barrera, 1981). Full face-processing abilities take several years to develop (de Haan, 2001).

Thus, face perception appears to be substantially organized at birth, yet only develops fully through postnatal experience. Low-level image processing in V1 appears to be able to explain some of the newborn face preferences that have been found, as shown recently in a computational model (Acerra, Burnod, and de Schonen, 2002), but many of them appear to require more face-specific circuitry (Bednar, 2002). To explain these more specific face preferences, previous theoretical models invoke separate visual processing mechanisms in newborn and older infants. For instance, Johnson and Morton (1991) proposed that infants are born with a basic subcortical system (CONSPEC) that serves only to detect and direct attention to face-like patterns in the periphery, perhaps using a simple three-dot template (two dots for the eyes and a third for the nose and mouth). A separate cortical system (CONLERN) begins to control behaviour after one month, and gradually develops more sophisticated face processing, through learning in central vision.

However, models like CONSPEC/CONLERN and that of Acerra *et al.* (2002) do not account for neonatal face learning. For example, an infant only a few days old prefers to look at its mother's face, relative to the face of a stranger (Bushnell, 2001; Pascalis *et al.*, 1995). Presumably this preference results from postnatal visual experience of the mother's face (Bushnell, 2001). Unlike the facial feature learning of CONLERN, this early learning has been thought to rely only on the external outline of the face, because the preference disappears when the face outline is masked (Pascalis *et al.*, 1995).

Accordingly, Johnson and Morton (1991) and subsequent authors have proposed adding face-outline learning to CONSPEC, or adding a third, separate subsystem for learning face outlines at birth (de Schonen, Mancini, and Liegeois, 1998; Simion *et al.*, 1998). However, recent studies suggest that newborns can also learn internal features (Slater, Bremner, Johnson, Sherwood, Hayes, and Brown, 2000). Such learning would seem to require a fourth subsystem, i.e. a learning system similar to CONLERN, but for the periphery, and operational at birth. Alternatively, CONLERN itself might be functional at birth, but in that case the model would need an additional mechanism to explain why schematic faces are no longer preferred in the periphery by two months of age. Our computational results will show that such increasingly complex models of early face processing are unnecessary. A single cortical learning system present at birth is sufficient to explain the experimental data, if the system is exposed to internally generated patterns broadly similar to faces.

Internally generated patterns

Our approach relies on the learning of internally generated patterns. Such spontaneous neural activity has recently been documented in many cortical and subcortical areas as they develop, including the retina, the visual cortex, the auditory system, and the spinal cord (reviewed in O'Donovan, 1999; Wong, 1999). In this work, we focus on two common types of spontaneous activity: retinal waves and ponto-geniculo-occipital (PGO) waves. Retinal waves are found in the developing retina before the eyes first open, and consist of large patches of coherent activity traveling across the retina, surrounded by inactive areas (Feller, Wellis, Stellwagen, Werblin, and Shatz, 1996; Maffei and Galli-Resta, 1990; Meister, Wong, Baylor, and Shatz, 1991). Retinal waves have been shown to be responsible for segregating the LGN into eye-specific layers before birth, indicating that internally generated activity patterns are crucial for the visual system to develop normally (Shatz, 1996; Stellwagen and Shatz, 2002).

PGO waves are activity patterns that are generated during rapid eye movement (REM) sleep. Developing embryos spend a large percentage of their time in a primitive form of REM sleep, which suggests that this state has a major role in development (Roffwarg, Muzio, and Dement, 1966). During REM sleep, PGO waves originate in the brain stem, then travel to the LGN, visual cortex, and many other brain areas (see Callaway, Lydic, Baghdoyan, and Hobson, 1987 for a review). PGO waves are strongly correlated with eye movements and with vivid visual imagery in dreams, suggesting that they activate the visual system as if they were visual inputs (Marks, Shaffery, Oksenberg, Speciale, and Roffwarg, 1995). PGO waves also elicit different distributions of activity in different species, and interrupting them has been shown to increase the influence of the environment on visual system development (Marks *et al.*, 1995).

Together, these findings suggest that spontaneous activity like retinal waves and PGO waves may act as training patterns for development (Jouvet, 1998). Retinal waves have been measured directly using optical imaging techniques, but it is not yet known what spatial patterns the PGO waves form (Rector, Poe, Redgrave, and Harper, 1997). How closely neonatal REM sleep and PGO waves are related to those of adults is also controversial (Dugovic and Turek, 2001; Jouvet, 1999). In any case, it is clear that there is substantial activity in the visual areas of newborns during sleep, and that it comes both from the eyes and from the brainstem. These assumptions about the spontaneous activity will be crucial for the experiments in this chapter.

Specifically, the orientation simulations model the effect of retinal waves on the primary visual cortex, based on the hypothesis that such patterns could drive the orientation selectivity and maps seen at birth. The face detection simulations explore how brainstem-generated patterns with higher level structure, such as groups of three similar dots, could drive the development of face detection and face preferences. The simulations show that much of what is known about how orientation maps develop and how newborns detect and learn faces can be explained by a single, general-purpose learning system, which learns from both internally generated patterns of activity and from the visual environment. We propose that this mechanism may be an efficient way to develop a robust system from a small amount of genetic information.

The HLISSOM model

The simulations are based on the HLISSOM model (Bednar, 2002). HLISSOM is a version of the LISSOM architecture (Sirosh and Miikkulainen, 1994) extended to process spontaneous activity, natural images, and faces. As shown in Figure 2.2, the model consists of a hierarchy of two-dimensional sheets of neural units modelling different areas of the visual system: a sheet of retinal photoreceptors, a sheet representing internally generated sources of activity (the PGO sheet), a pair of sheets of LGN units (ON-centre and OFF-centre),

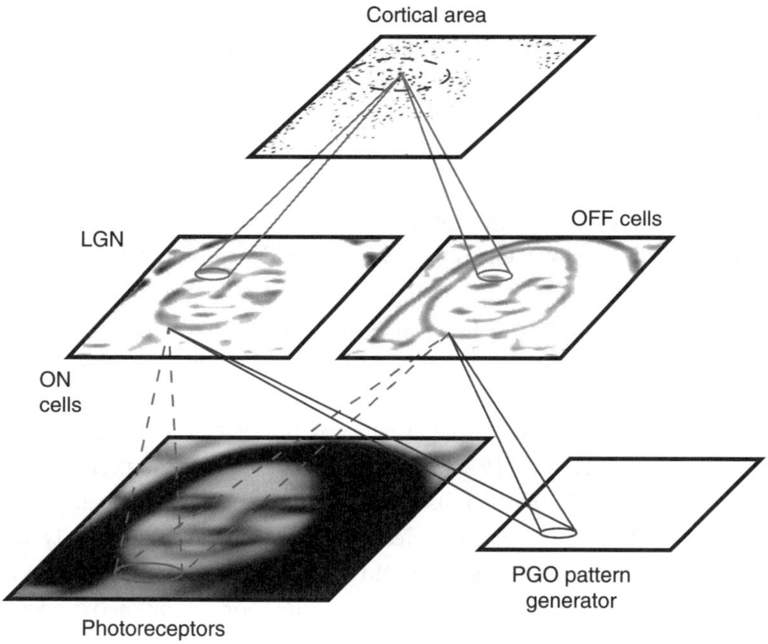

Fig. 2.2 HLISSOM model. The model consists of a hierarchy of sheets of neural units, modeling the infant visual pathway. Input such as grayscale images or spontaneous activity patterns is drawn on the photoreceptor array or the PGO array, which represents sources of spontaneous activity. Each higher level unit computes its response from the activity in its receptive fields (RFs); the RFs of one unit in each map are shown. The cortical area corresponds to either V1, for orientation map simulations, or the FSA, for face processing simulations. Cortical units have receptive fields both on previous levels (afferent connections) and on the same level (lateral connections). V1 simulations use a cortical area with small receptive fields to model low-level orientation-selective cells, while FSA simulations use large receptive fields to model face-selective cells. V1 and the FSA can also be arranged hierarchically into the same model, which gives similar results but is more complex to simulate and analyze (Bednar, 2002). Photograph copyright 1998, James A. Bednar.

and a sheet of cortical units ('neurons'). As in the cortex, the overall arrangement of areas is fixed in HLISSOM, but the smaller-scale organization of cortical maps and connections is learned from input activity.

The photoreceptor and PGO input sheets act like grayscale bitmap images, i.e., arrays of numeric values (as shown in Figure 2.2). The ON-centre LGN units respond to patches on the input sheets that are brighter than their surrounds, and the OFF-centre units respond to patches darker than their surrounds. Thus the LGN sheets model the edge detection performed by the LGN and the retinal ganglia, highlighting boundaries and edges in the input. The cortical area corresponds to human V1 for the orientation simulations and a higher level area called the face-selective area (FSA) for the face-processing simulations. The FSA represents the first region in the ventral processing pathway that has receptive fields large enough to span a human face at close range.

For simplicity, the simulations in this paper use only a single cortical area, either V1 or the FSA, but results are similar with a more complex model where the FSA receives input through a lower level V1 region (Bednar 2002). Each cortical neuron in V1 or the FSA corresponds to a vertical column of cells through the six anatomical layers of the cortex. These neurons learn to represent the important features of the LGN input, based on experience with a series of input patterns. After such self-organization, the cortical neurons respond to particular orientations (for the orientation simulations) or objects (for the face simulations).

The rest of this section reviews the process of self-organization in more detail; see Miikkulainen, Bednar *et al.* (2005) for a complete description. Each LGN unit computes its response as a product between its weight vector and its receptive fields on the photoreceptors and PGO layers (see Equation 1 in Box 2.1). The LGN weight vectors are fixed radial Difference-of-Gaussian (DoG) functions. These sombrero-shaped RFs ensure that the ON LGN units respond to light areas surrounded by dark (by subtracting a broad Gaussian from a narrow one) and that OFF LGN units respond to dark areas surrounded by light (by subtracting a narrow Gaussian from a broad one).

Each cortical neuron computes its initial response from its weight vector and its input like an LGN unit does, except that the cortical RFs are on the ON and OFF LGN sheets. After the initial response, the cortical activity settles through short-range excitatory and long-range inhibitory lateral interaction (Equation 2 in Box 2.1). The contribution from the lateral connections is computed as a similar product between a weight vector and a receptive field, except that the activity of the units in the RF is taken from the preceding settling step. The cortical activity pattern starts out diffuse, but within a few settling iterations, it

Box 2.1 HLISSOM model equations

LGN activation

$$\eta_{ij} = \sigma(\Sigma_{\rho}\gamma_{\rho}\Sigma_{ab}X_{\rho ab}\omega_{ij,\rho ab}), \quad (1)$$

where:

(i, j) specify an LGN unit
η_{ij} is the response of that unit
ρ iterates over the RFs on the input sheets (the photoreceptors and the PGO layer)
σ is a piecewise linear sigmoid activation function
$X_{\rho ab}$ is the activation of input unit (a, b) on sheet ρ
$\omega_{ij,\rho ab}$ is the corresponding weight value

The initial (afferent) response of cortical (V1 or FSA) units is computed using the same equation, but with ρ iterating over the ON and OFF LGN sheets.

Cortical activation

$$\eta_{ij}\ (s) = \sigma(\Sigma_{\rho}\gamma_{\rho}\Sigma_{ab}X_{\rho ab}(s-1)\omega_{ij,\rho ab}) \quad (2)$$

where:

ρ iterates over the four RFs to cortical unit (i, j)
(the ON and OFF LGN RFs and the lateral excitatory and inhibitory weights)
$X_{\rho ab}(s-1)$ is the activation of input unit (a, b) during the previous settling step.

Cortical learning

$$\omega_{ij,\rho ab}(f+1) = \omega_{ij,\rho ab}, (f) + \alpha_{\rho}\eta_{ij}X_{\rho ab}$$

$$\Sigma_{ab}[\omega_{ij,\rho ab}(f) + \alpha_{\rho}\eta_{ij}X_{\rho ab}] \quad (3)$$

where:

η_{ij} is the activity of neuron (i, j) in the final activity bubble
$\omega_{ij,\rho ab}\ (f)$ is the connection weight from the previous iteration
α is the learning rate for each type of connection
$X_{\rho ab}$ is the presynaptic activity.

converges into a small number of stable focused patches of activity, or activity bubbles (as in Figure 2.2).

After the activity has settled, the connection weights of each cortical neuron are modified. All cortical weights (including the lateral weights) adapt according to the Hebb rule, normalized so that the sum of the weights from each sheet is constant for each neuron (Equation 3 in Box 2.1).

The larger the product of the pre- and postsynaptic activity, the larger the weight change. The normalization ensures that weights do not grow without bound, and approximates homeostatic regulation processes that keep the overall excitability of a neuron constant (reviewed in Turrigiano, 1999). At long distances, few neurons have correlated activity and therefore most long-range connections eventually become weak. The weakest connections are eliminated periodically, resulting in patchy lateral connectivity similar to that observed in the visual cortex.

For the orientation map experiments reported in this paper, the ON-centre and OFF-centre LGN sheets had 36×36 units, and received input from a 54×54 array. For the face-processing experiments, the ON and OFF sheets were 74×74, and the input array was 170×170 units. The V1 orientation map contained 96×96 neurons, and the FSA contained 24×24 neurons. Initially, the afferent weights of the cortical neurons were random, and the lateral weights had a smooth circular Gaussian profile. The afferent RFs were 13 units wide in V1, and 25 units wide in the FSA (larger because higher level cortical areas have larger RFs). The rest of the parameters are described in Bednar (2002).

Orientation map experiments

To study how prenatal and postnatal learning together produce the adult orientation map, development was simulated in two phases. In the prenatal phase of 1000 input presentations, input patterns consisted of uniform random noise combined with a large 'disk' of activity representing a local patch of highly responding neurons. Figure 2.3(b) shows an example of this type of pattern. These patterns were chosen to match retinal waves, which are the best-characterized source of spontaneous activity in early development. However, they can also represent any other spontaneous activity that includes large patches that are spatially coherent. The 9000 iterations constituting the postnatal phase used images of natural objects and scenes, modelling visual experience (see Figures 2.3 d–f).

This network first develops an orientation map through spontaneous activity, then gradually refines the map based on experience with natural images,

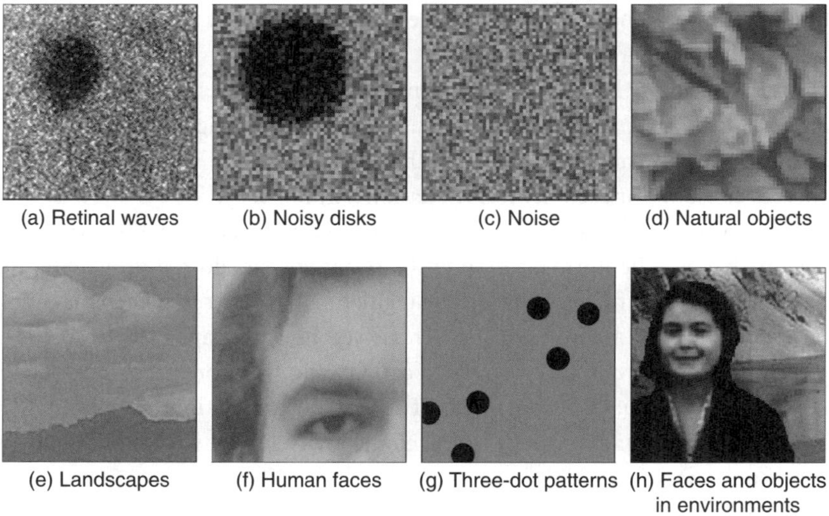

Fig. 2.3 Training patterns. Plot (a) shows an example of a retinal wave in the ferret, with active areas appearing dark. (a reprinted with permission from Feller *et al.*, (1996), *Science*, 272, 1182–1187, copyright 1996 American Association for the Advancement of Science.) The prenatal orientation map simulations are based on an approximation to these patterns (b). As a control, we also tested whether initial orientation maps can develop from uniform random noise alone (c). Postnatal orientation map simulations use natural images (d) or specific types of stimuli, such as natural landscapes (e) or closeups of human faces (f). Prenatal face processing simulations are similar but use more structured inputs, usually without noise to reduce computational requirements (g). Postnatal face processing simulations use photographs of people or objects presented at random locations in front of photographs of natural scenes.

without changing the overall shape of the map (Figure 2.4). The final orientation preference distribution is also biased towards horizontal and vertical, as seen in ferret visual cortex (Coppola *et al.*, 1998). These results are the main novel component of the orientation map simulations. To our knowledge, only one other computational model has been tested in this way (Burger and Lang, 1999). Unlike in animals or in HLISSOM, postnatal training in that model entirely rearranged the map, perhaps because their prenatal training patterns did not contain oriented edges as their postnatal patterns did. If the Burger and Lang model is tested with patterns like retinal waves, it may also show the smooth progression from prenatal to postnatal organizations seen in HLISSOM.

Further HLISSOM experiments explored how different pattern types can change the map organization (Bednar, 2002). These three experiments act as controls for the main results, helping to determine what features of the model

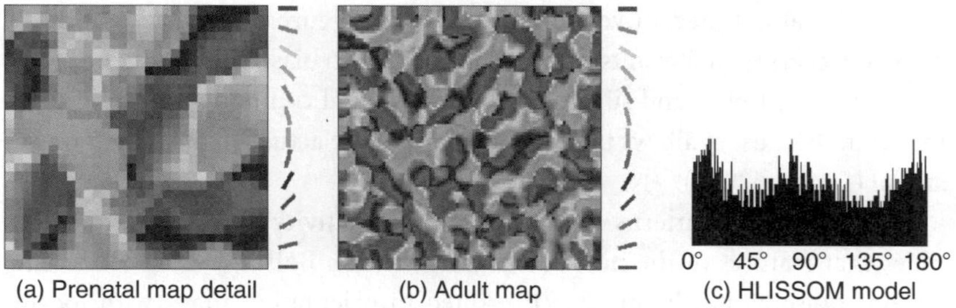

Fig. 2.4 HLISSOM model orientation maps. By iteration 1000, the end of prenatal training, the HLISSOM map was a good match to newborn maps (compare the central area plotted in a with Figure 2.1a). Further training on 9000 natural image patches results in an adult-like orientation map (compare b with Figure 2.1b). The statistics of the orientation preferences of the adult map reflect the horizontal and vertical biases of the training images, causing the map to devote more area to horizontal and vertical (compare c with Figure 2.1c).

are crucial. First, different postnatal training conditions were tested, to see how different environments can change the outcome. The maps indeed reflected the skewed data sets. For the Landscapes dataset (Figure 2.3e), containing many horizontal contours, the orientation map and histogram became biased towards horizontal (Figure 2.5b). For the Faces dataset (Figure 2.3f), more

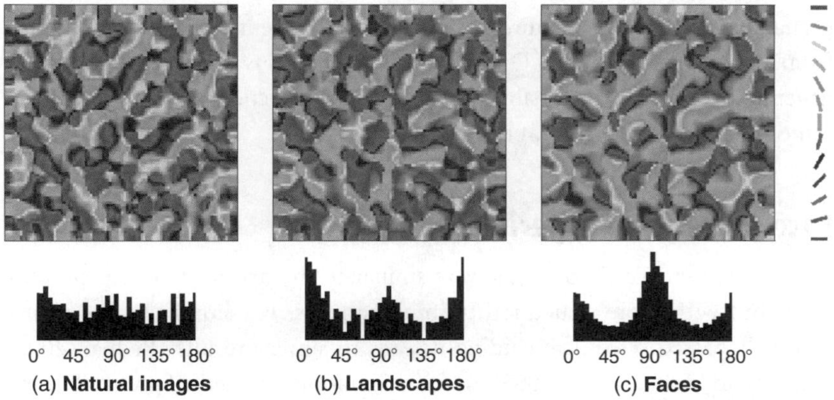

Fig. 2.5 Effect of different postnatal training environments. Three copies of the same prenatally organized map were trained postnatally with different image datasets: Nature (as shown in Figure 2.4c), Landscapes (a collection of nature scenes including many strong horizontal contours), and Faces (a collection of frontal views of faces). Although the resulting maps all have similar large-scale organization determined by the prenatal training, the number of neurons selective for each orientation reflects the distribution of postnatal training patterns. Thus prenatal training provides the basic structure, and postnatal training helps the map represent the most common patterns in the environment.

neurons became tuned to vertical orientations (Figure 2.5c). These findings replicate experimental results from animals raised in biased environments (e.g. Sengpiel *et al.*, 1999), and suggest that the postnatal component of training is important because it allows the map to adapt to the actual environment of the animal (Bednar, 2002).

Second, different patterns of prenatal training activity were tested, to determine what features of the noisy disks were crucial. Both the disks alone and random noise alone (Figure 2.3c) resulted in orientation maps with locally coherent patches of orientation-selective neurons. However, the receptive fields in the disks case were much more selective than in typical newborn neurons, indicating that including noise was crucial for obtaining realistic receptive fields. Yet for uniform random noise alone, the receptive fields were poorly selective, and primarily included receptive field types rarely reported in experimental studies of V1, such as crosses (Bednar, 2002). The orientation map was also less well organized than typical newborn maps. These results suggest that both spatially uncorrelated noise and spatially correlated activity are necessary to develop a normal prenatal map.

Third, an initially random map was trained on natural images alone, to test whether the prenatal training was crucial for developing orientation maps and selectivity. Although the details of the final map organization differed from the normal case, the overall properties were very similar (Bednar, 2002). Thus prenatal training is not required for a functional orientation map to develop from natural image input. Overall, these simulations show how internally generated patterns can explain orientation map function at birth, while the same system later learns from real visual images.

Face preference experiments

The face preference simulations were similar to the orientation map simulations, but with more structured training patterns. For simplicity, the simulations bypassed V1, but including it leads to similar results (Bednar, 2002; Bednar and Miikkulainen, 2003). As for orientation, a variety of prenatal training conditions were simulated, to determine how different training patterns can lead to face preferences. Most simulations used the three-dot patterns shown in Figure 2.3g, presented at randomly chosen positions and random near-vertical orientations. For control purposes, networks were also trained using other internally generated patterns, and an untrained (naive) network was also tested. Postnatally, the training regime represented gradual learning of specific individuals and objects seen against a variety of different backgrounds. For each postnatal iteration, one of six face images or six object images was chosen randomly, and presented in front of a randomly chosen natural scene background (Bednar, 2002) (see Figure 2.3h). The main goals of these simulations were to test whether prenatal learning could account for human performance at birth, and whether this learning shapes postnatal learning and the final self-organized architecture.

After prenatal training, and at intervals throughout postnatal training, the network was tested using schematic images previously tested with newborns, and with photographs of faces. In order to compare the neural activity levels in the model to the patterns preferred by babies, we assumed that an infant pays more attention to the stimuli that are most effective at activating its visual processing system, focusing on the highest level activated (Bednar, 2002). Patterns activating the FSA will be preferred over those activating only lower areas, and two patterns that both activate the FSA will be ranked by their total FSA activity. Under these assumptions, it is possible to compare the model results with psychological data from human infants.

When trained on three-dot patterns at various locations and orientations, FSA neurons developed receptive fields preferring upright, triangular arrangements of three dots (Figure 2.6a, row RF 10000). These RFs cause the network to respond to facelike schematic images (Figure 2.6), with total activity levels that rank the patterns in the order preferred by infants (Goren *et al.*, 1975; Johnson and Morton, 1991). When tested on 18 schematic patterns from newborn studies (Goren *et al.*, 1975; Johnson and Morton, 1991; Simion *et al.*, 1998; Valenza, Simion, Cassia, and Umiltà, 1996), the full version of the model (including both V1 and the FSA) ranked them in the same preference order for all 22 of the statistically significant preferences found in newborns (Bednar, 2002; Bednar and Miikkulainen, 2003). These results demonstrate that a network exposed to three-dot patterns is sufficient to explain the experimental results with newborns tested with schematic stimuli.

Although schematic patterns are usually used in newborn studies for practical reasons, it is important to show that the HLISSOM model also exhibits preferences for real faces compared to other photographs. Given a database of 150 top-lit images of adult males, the model responded in the correct location (the centre of the face) to 88 per cent of the images (Bednar and Miikkulainen, 2003). Conversely, it responded to only 4 per cent of the images in a database of natural scenes. Overall, the FSA responded to most top-lit human faces, signaling their location in the visual field. It did not respond to most other stimuli, except when they contained accidental three-dot patterns. The model predicts that human newborns will have a similar pattern of responses in the face-selective cortical regions.

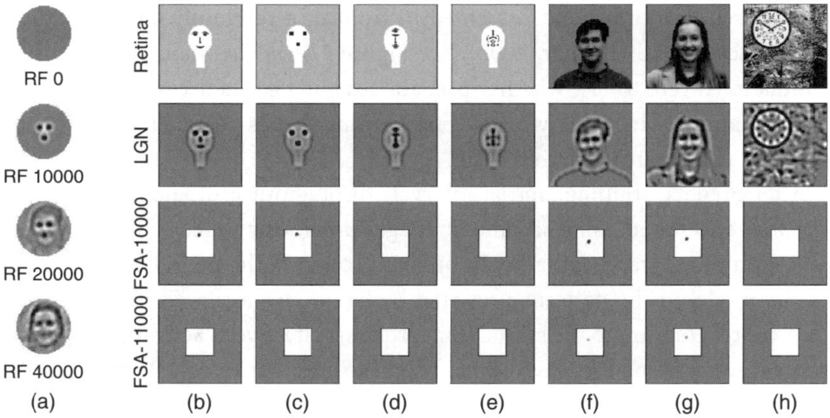

Fig. 2.6 Model RFs and responses. The plots in (a) show the RF for a typical neuron in the FSA at different stages of training; the pattern shown is the LGN activity that would most excite this neuron. At first the neuron is unselective (RF 0), but after prenatal exposure to many three-dot patterns it comes to prefer them (RF 10000). The RF then gradually becomes more selective for real faces through postnatal experience (RF 20000 and RF 40000). The remaining plots show how such neurons respond to test images. The top row shows four schematic patterns (Johnson and Morton, 1991) and two photographic stimuli (Rowley *et al.*, 1998). Below each is the LGN response to that image (second row), the FSA response after prenatal training on three-dot patterns (third row), and the FSA response after postnatal training on real faces (bottom row). The 'neonatal' FSA (row FSA-10000) responds in the location of face-like schematic and real stimuli, and rarely responds to other images. As in two-month-olds, the response to schematic patterns eventually disappears when the network is trained on real images (row FSA-11000), but the responses to real faces remain (f–g). For images without faces, such as the training pattern shown in (h), the network responds only when there are accidental three-dot patterns.

How crucial is the specific training pattern for the prenatal preferences? In addition to the three-dot pattern proposed by Johnson and Morton (1991), we tested a variety of other prenatal training pattern shapes (Bednar, 2002; Bednar and Miikkulainen, 2003). For all training patterns, test faces that matched the training pattern size gave higher responses than did other natural images. However, responses to the schematic patterns were very different for networks trained with different patterns. Of all training patterns tested, the three-dot pattern provided the best selectivity for faces, and a better fit to the experimental data with schematics.

Importantly, not all patterns result in face preferences. For example, if only a single low-level feature like eye size is matched between the training and testing images, the HLISSOM model does not prefer faces. These results show that the pattern shape is important, but that a variety of shapes lead to qualitatively similar results.

What effect does the prenatal training have on postnatal learning? Trained with real faces in real images, most units in both the prenatally trained and naive networks developed RFs that are averages (i.e. prototypes) of faces and hair outlines (Bednar and Miikkulainen, 2002). However, the time course of learning and the final result differed. RFs in the prenatally trained network become gradually more selective for faces. Postnatal self-organization in the naive network is less regular, and the final RFs are usually less selective. Overall, prenatal training appears to bias postnatal learning towards biologically relevant stimuli, i.e. faces (as suggested by Johnson and Morton, 1991).

The bottom row of Figure 2.6 shows that HLISSOM gradually responds less to schematic faces, just as infants no longer prefer such faces in the periphery after one month (Johnson *et al.*, 1991). In HLISSOM, the loss of face preferences is caused by afferent weight normalization. As the FSA neurons in HLISSOM learn the hair and face outlines typically associated with real faces, the connections to the internal features necessarily become weaker. As a result, the response to face-like schematic patterns decreases, because those patterns match only the internal features.

Eventually, the response to the schematic patterns drops below the fixed activation threshold. At that point, the model does not respond more strongly to schematic faces than to other schematics. In a sense, the FSA has learned that real faces typically have both inner and outer features, and does not respond when either type of feature is absent or a poor match to real faces.

Interestingly, the FSA neurons continue to respond to real faces (as opposed to schematics) throughout postnatal learning (Figure 2.6 f–g). Thus the model provides a clear prediction that even if infants lose their preference for schematics over time, they will continue to prefer realistic faces compared with other similar images.

When one face (i.e. the mother) appears most often, the FSA response to that face becomes stronger than to a similar stranger (Figure 2.7) (Bednar and Miikkulainen, 2002). This result replicates the mother preference found in infants a few days old (Bushnell, 2001; Pascalis *et al.*, 1995). Interestingly, the model no longer prefers the mother when her hair outline is masked, which is consistent with Pascalis *et al.*'s claim that newborns learn outlines only. However, Pascalis *et al.* (1995) did not test the crucial converse condition, i.e. whether newborns respond when the facial features are masked, leaving only the outlines. As for the internal features alone, the model does not respond to the head and hair outline alone. Thus the HLISSOM model's learning of faces is clearly not restricted to the outline.

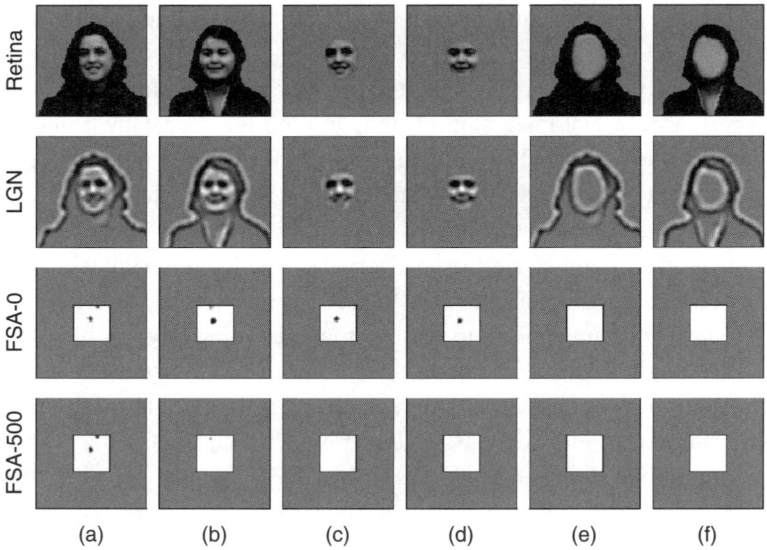

Fig. 2.7 Mother preferences depend on both internal and external features. Initially, the prenatally trained FSA responds well to both women above (a–b; FSA-0), with no significant difference ($p = 0.28$). The response is primarily due to the internal facial features (c–d; FSA-0), although there are some spurious three-dot responses due to alignment of the hair with the eyes (a–b; top of FSA-0). Designating image (a) as the mother, we presented it in 25 per cent of the postnatal learning iterations. This ratio is taken from Bushnell (2001), who found that newborns look at their mother's face for an average of about one-fourth their time awake over the first few days. Image (b), the stranger, was not presented at all during training. After 500 postnatal iterations, the response to the mother is significantly greater than to face (b) ($p = 0.001$). This result replicates the mother preference found by Pascalis *et al.*, (1995) in infants 3–9 days old. The same results are found in the counterbalancing condition—when trained on face (b) as the mother, (b) becomes preferred ($p = 0.002$; not shown). After training with real faces, there is no longer any FSA response to the facial features alone (c–d), which replicates Pascalis *et al.*'s (1995) finding that newborns no longer preferred their mother when her face outline was covered. Yet contra Pascalis *et al.*, (1995), we cannot conclude that what has been learned 'has to do with the outer rather than the inner features of the face', because no preference is found for the face outline alone either (e–f). These results suggest that newborn face learning is holistic, and includes all features associated with faces.

In the model, the response is lower with either type of masking because of holistic learning of all of the features of real faces. As real faces are learned, the afferent weight normalization ensures that neurons respond only to patterns that are a good overall match to all of the weights, not simply matching on a few features. Many authors have argued that adults learn faces holistically (e.g. Farah, Wilson, Drain, and Tanaka, 1998). The simulation results suggest that newborns may learn faces in the same way, and thus that they will not prefer their mother when her internal facial features are masked. Overall, the face preference experiments provide a detailed computational simulation of how preferences can be constructed prenatally and refined postnatally, and provide concrete predictions for future experiments.

Discussion and future work

The results presented in this chapter show how environmental and genetic factors can interact to produce selectivity for both high-level and low-level visual features. Both environmental and genetic factors are crucial for explaining the experimental data, and both are incorporated using the same general-purpose learning algorithm. In each case, the prenatal organization provides structure at birth, while postnatal learning allows these initial preferences to be refined to match the actual visual patterns in the environment.

The orientation simulations show that the development of low-level features like orientation maps is robust, i.e. maps can develop from a wide range of possible input patterns. Spontaneous activity with actual oriented edges, like retinal waves, may help ensure that maps develop normally in a wider range of possible environments. Prenatal training may also allow V1 to mature more quickly, which could be important for organizing higher levels of the visual system.

The face preference simulations show that prenatal learning of spontaneous activity and postnatal learning of real faces can together account for newborn face preferences, neonatal face learning, and longer term development of face detection. The results suggest simple but novel explanations for why newborns prefer schematic faces, why newborn learning appears to depend on the face outline, and why the response to schematic faces decreases over time. They also suggest that internally generated patterns allow the genome to steer development towards biologically relevant processing, making subsequent learning more robust.

The orientation and face preference results depend primarily on having a model that learns representations for common patterns on the input, plus having prenatal patterns that share some structure with those experienced postnatally. Thus we expect that similar results could be found with other realistic models of visual processing, such as those of Burger and Lang (1999) for orientation and Rolls and Milward (2000) for face recognition. In each case, the crucial assumptions are (1) the type of patterns that are present before visual experience, (2) the overall arrangement of visual areas, and (3) the 'hard-wired' processing in the early visual areas. In future work, it will be important to measure the actual activity patterns in developing animals and humans, and understand how these patterns are generated in the brainstem and other areas.

Brain imaging in young infants can help determine whether face-selective responses in newborns are subcortical, as in the CONSPEC/CONLERN model, or cortical, as in the HLISSOM model. Even before newborn imaging is practical, the predictions of the face processing model can be tested behaviourally with infant experiments. For example, HLISSOM predicts that over the first two months the response to real faces in the periphery should continue, even as response to schematics diminishes, and the mother preference of newborns should disappear when the facial features are masked. These experiments will help clarify the roles of internally generated activity in the development of the cortex.

Conclusion

The HLISSOM model shows how the interaction between environmental and genetic factors can explain many features of orientation map and face processing develoment. In the model, the genetic factors take several forms: (1) constraints on the overall arrangement of visual areas and connectivity between areas, (2) prespecification of processing in early visual areas such as the retina and the LGN, and (3) genetically determined activity patterns generated during early development. Given these constraints and initial patterns, the model shows how a functional visual system can develop reliably, even without detailed genetic specification of neural connectivity patterns or of specific cortical representations.

At the same time, the specific representations that do develop depend strongly on interaction with the environment. Without environmental input, the model representations remain crude and relatively unselective, unlike neurons in adult cortical areas. Moreover, the specific properties of the environmental input are crucial: a preponderance of certain orientations or objects will resultin representations biased towards common stimuli, such as horizontal contours or the mother's face. This adaptation helps ensure that appropriate representations will develop.

In the model, the environmental interactions were simulated using static photographic images, but in general an organism will develop through interaction with a much richer environment, including moving patterns, associations between different sensory and motor modalities, and associations between sensory patterns and rewards or pain. We expect that these additional consequences of embodiment or situatedness will be crucial for explaining the full range of cortical development. For instance, experience with faces from a wide range of orientations, in different lighting conditions, and with different hairstyles is likely to be necessary for the development of adult-like face recognition that is invariant to these factors. Future work will explore how these more realistic types of embodiment can be studied computationally.

In summary, the specific simulations described above demonstrate how environmental and genetic factors can together explain much of how orientation and face processing develop. In both cases, the overall approach is the same, with initial orientation and face selectivity developing prenatally and being refined postnatally to match the visual environment. The modelling results provide clear predictions for future animal and infant experiments, and they act as examples of a new way of understanding how sensory systems can be constructed.

Acknowledgements

This research was supported in part by the National Science Foundation under grants IRI-9309273 and IIS-9811478, and by the National Institute of Mental Health under Human Brain Project grant R01-MH66991-01. Software, demonstrations and related publications are available at computationalmaps.org and topographica.org.

Glossary

Piecewise-linear sigmoid activation function A neural network unit generates a real-valued output (its activation) as a function of the values of its inputs. A sigmoid function is an S-shaped curve that produces little activation for weak inputs, nearly linear activation for intermediate values, and saturates quickly for high values. A piecewise-linear sigmoid is an approximation to the smooth sigmoid curve, using linear functions for the low, medium, and high regimes, which is more computationally efficient to simulate.

Topographic neural maps Cortical sensory areas are organized in terms of topographic maps of neurons. For instance, most visual areas are retinotopic, mapping from input organized spatially in two dimensions like the retina, into another retinotopic representation. Other sensory and motor areas also have two-dimensional topographic organization.

Ventral processing pathway Ungerleider and Mishkin (1982) proposed that the organization of the visual system could be understood in terms of a ventral stream and a dorsal stream, also called the what and the where pathways. Areas in the ventral stream, such as IT (inferotemporal cortex), are thought to be concerned with object identity. Those in the dorsal stream, such as posterior parietal cortex, are thought to be concerned primarily with precise object location and the detection of motion. The two streams share early cortical areas such as V1 and V2.

Hebb rule Donald Hebb (1949) proposed a neural network learning rule that is now called the Hebb rule or the Hebbian rule. He suggested that neurons that often activate at the same time should become more strongly connected, an idea that has since received wide experimental and theoretical support. A variety of mathematical equations can be characterized as Hebbian, as long as they result in connections being strengthened between coactive neurons.

References

Acerra F, Burnod Y, and de Schonen S (2002). Modelling aspects of face processing in early infancy. *Developmental Science*, 5(1), 98–117.

Bednar JA (2002). *Learning to see: Genetic and environmental influences on visual development.* PhD thesis, Department of Computer Sciences, The University of Texas at Austin. Technical Report AI-TR-02–294.

Bednar JA and Miikkulainen R (2002). Neonatal learning of faces: Interactions between genetic and environmental inputs. In W Gray and C Schunn (eds) *Proceedings of the 24th Annual Conference of the Cognitive Science Society*, pp. 107–112. Mahwah, NJ: Erlbaum.

Bednar JA and Miikkulainen R (2003). Learning innate face preferences. *Neural Computation*, 15(7), 1525–1557.

Blakemore C and Cooper GF (1970). Development of the brain depends on the visual environment. *Nature*, 228, 477–478.

Blakemore C and van Sluyters RC (1975). Innate and environmental factors in the development of the kitten's visual cortex. *Journal of Physiology (London)*, 248, 663–716.

Blasdel GG (1992). Orientation selectivity, preference, and continuity in monkey striate cortex. *Journal of Neuroscience*, 12, 3139–3161.

Burger T and Lang EW (1999). An incremental Hebbian learning model of the primary visual cortex with lateral plasticity and real input patterns. *Zeitschrift für Naturforschung C–A Journal of Biosciences*, 54, 128–140.

Bushnell IWR (2001). Mother's face recognition in newborn infants: Learning and memory. *Infant and Child Development*, 10(1/2), 67–74.

Callaway CW, Lydic R, Baghdoyan HA, and Hobson JA (1987). Pontogeniculooccipital waves: Spontaneous visual system activity during rapid eye movement sleep. *Cellular and Molecular Neurobiology*, 7(2), 105–149.

Chapman B and Bonhoeffer T (1998). Overrepresentation of horizontal and vertical orientation preferences in developing ferret area 17. *Proceedings of the National Academy of Sciences, USA*, 95, 2609–2614.

Chapman B and Stryker MP (1993). Development of orientation selectivity in ferret primary visual cortex and effects of deprivation. *Journal of Neuroscience*, 13(12), 5251–5262.

Chapman B, Stryker MP, and Bonhoeffer T (1996). Development of orientation preference maps in ferret primary visual cortex. *Journal of Neuroscience*, 16(20), 6443–6453.

Coppola DM, White LE, Fitzpatrick D, and Purves D (1998). Unequal representation of cardinal and oblique contours in ferret visual cortex. *Proceedings of the National Academy of Sciences, USA*, 95(5), 2621–2623.

Crair MC, Gillespie DC, and Stryker MP (1998). The role of visual experience in the development of columns in cat visual cortex. *Science*, 279, 566–570.

de Haan M (2001). The neuropsychology of face processing during infancy and childhood. In CA Nelson and M Luciana (eds) *Handbook of developmental cognitive neuroscience*, pp. 381–398. Cambridge, MA: MIT Press.

de Schonen S, Mancini J, and Liegeois F (1998). About functional cortical specialization: The development of face recognition. In F Simion and G Butterworth (eds) *The development of sensory, motor and cognitive capacities in early infancy: From perception to cognition*, pp. 103–120. Hove: Psychology Press.

Diamond S (1974). Four hundred years of instinct controversy. *Behavior Genetics*, 4, 237–252.

Dugovic C and Turek FW (2001). Similar genetic mechanisms may underlie sleep-wake states in neonatal and adult rats. *Neuroreport*, 12(14), 3085–3089.

Farah MJ, Wilson KD, Drain M, and Tanaka JN (1998). What is 'special' about face perception? *Psychological Review*, 105(3), 482–498.

Feller MB, Wellis DP, Stellwagen D, Werblin FS, and Shatz CJ (1996). Requirement for cholinergic synaptic transmission in the propagation of spontaneous retinal waves. *Science*, 272, 1182–1187.

Gödecke I, Kim DS, Bonhoeffer T, and Singer W (1997). Development of orientation preference maps in area 18 of kitten visual cortex. *European Journal of Neuroscience*, 9(8), 1754–1762.

Goren CC, Sarty M, and Wu PY (1975). Visual following and pattern discrimination of face-like stimuli by newborn infants. *Pediatrics*, 56(4), 544–549.

Hebb DO (1949). *The organization of behavior: A neuropsychological theory*. New York: Wiley.

Hubel DH and Wiesel TN (1962). Receptive fields, binocular interaction and functional architecture in the cat's visual cortex. *Journal of Physiology (London)*, 160, 106–154.

Hubel DH and Wiesel TN (1968). Receptive fields and functional architecture of monkey striate cortex. *Journal of Physiology (London)*, 195, 215–243.

Johnson MH, Dziurawiec S, Ellis H, and Morton J (1991). Newborns' preferential tracking of face-like stimuli and its subsequent decline. *Cognition*, 40, 1–19.

Johnson MH and Morton J (1991). *Biology and cognitive development: The case of face recognition*. Oxford and New York: Blackwell.

Jouvet M (1998). Paradoxical sleep as a programming system. *Journal of Sleep Research*, 7(Suppl. 1), 1–5.

Jouvet M (1999). *The paradox of sleep: The story of dreaming*. Cambridge, MA: MIT Press.

Kandel ER, Schwartz JH, and Jessell TM (1991). *Principles of neural science*, 3rd edn. New York: Elsevier.

Kanwisher N, McDermott J, and Chun MM (1997). The fusiform face area: A module in human extrastriate cortex specialized for face perception. *Journal of Neuroscience*, 17(11), 4302–4311.

Linsker R (1986). From basic network principles to neural architecture: Emergence of orientation columns. *Proceedings of the National Academy of Sciences, USA*, 83, 8779–8783.

Maffei L and Galli-Resta L (1990). Correlation in the discharges of neighboring rat retinal ganglion cells during prenatal life. *Proceedings of the National Academy of Sciences, USA*, 87, 2861–2864.

Marks GA, Shaffery JP, Oksenberg A, Speciale SG, and Roffwarg HP (1995). A functional role for REM sleep in brain maturation. *Behavioural Brain Research*, 69, 1–11.

Maurer D and Barrera M (1981). Infants' perception of natural and distorted arrangements of a schematic face. *Child Development*, 52(1), 196–202.

Meister M, Wong ROL, Baylor DA, and Shatz CJ (1991). Synchronous bursts of action-potentials in the ganglion cells of the developing mammalian retina. *Science*, 252, 939–943.

Miikkulainen R, Bednar JA, Choe Y, and Sirosh S (2005). *Computational Maps in the Visual Cortex*, Springer 2005.

Miyashita-Lin EM, Hevner R, Wassarman KM, Martinez S, and Rubenstein JL (1999). Early neocortical regionalization in the absence of thalamic innervation. *Science*, 285(5429), 906–909.

Movshon JA and van Sluyters RC (1981). Visual neural development. *Annual Review of Psychology*, 32, 477–522.

O'Donovan MJ (1999). The origin of spontaneous activity in developing networks of the vertebrate nervous system. *Current Opinion in Neurobiology*, 9, 94–104.

Pascalis O, de Schonen S, Morton J, Deruelle C, and Fabre-Grenet M (1995). Mother's face recognition by neonates: A replication and an extension. *Infant Behavior and Development*, 18, 79–85.

Rakic P (1988). Specification of cerebral cortical areas. *Science*, 241, 170–176.

Rector DM, Poe GR, Redgrave P, and Harper RM (1997). A miniature CCD video camera for high-sensitivity light measurements in freely behaving animals. *Journal of Neuroscience Methods*, 78(1–2), 85–91.

Roffwarg HP, Muzio JN, and Dement WC (1966). Ontogenetic development of the human sleep–dream cycle. *Science*, 152, 604–619.

Rolls ET and Milward T (2000). A model of invariant object recognition in the visual system: Learning rules, activation functions, lateral inhibition, and information-based performance measures. *Neural Computation*, 12(11), 2547–2572.

Rowley HA, Baluja S, and Kanade T (1998). Neural network-based face detection. *IEEE Transactions on Pattern Analysis and Machine Intelligence*, 20(1), 23–38.

Sengpiel F, Stawinski P, and Bonhoeffer T (1999). Influence of experience on orientation maps in cat visual cortex. *Nature Neuroscience*, 2(8), 727–732.

Shatz CJ (1996). Emergence of order in visual system development. *Proceedings of the National Academy of Sciences, USA*, 93, 602–608.

Simion F, Valenza E, and Umiltà C (1998). Mechanisms underlying face preference at birth. In F Simion and G Butterworth (eds) *The development of sensory, motor and cognitive capacities in early infancy: From perception to cognition*, pp. 87–102. Hove: Psychology Press.

Sirosh J and Miikkulainen R (1994). Cooperative self-organization of afferent and lateral connections in cortical maps. *Biological Cybernetics*, 71, 66–78.

Slater A, Bremner G, Johnson SP, Sherwood P, Hayes R, and Brown E (2000). Newborn infants' preference for attractive faces: The role of internal and external facial features. *Infancy*, 1(2), 265–274.

Slater A and Johnson SP (1998). Visual sensory and perceptual abilities of the newborn: Beyond the blooming, buzzing confusion. In F Simion and G Butterworth (eds) (1998). *The development of sensory, motor and cognitive capacities in early infancy: From perception to cognition*, pp. 121–142. Hove: Psychology Press.

Slater A, Morison V, and Somers M (1988). Orientation discrimination and cortical function in the human newborn. *Perception*, 17, 597–602.

Stellwagen D and Shatz CJ (2002). An instructive role for retinal waves in the development of retinogeniculate connectivity. *Neuron*, 33(3), 357–367.

Sur M, Angelucci A, and Sharma J (1999). Rewiring cortex: The role of patterned activity in development and plasticity of neocortical circuits. *Journal of Neurobiology*, 41, 33–43.

Sur M and Leamey CA (2001). Development and plasticity of cortical areas and networks. *Nature Reviews Neuroscience*, 2(4), 251–262.

Swindale NV (1996). The development of topography in the visual cortex: A review of models. *Network—Computation in Neural Systems*, 7, 161–247.

Switkes E, Mayer MJ, and Sloan JA (1978). Spatial frequency analysis of the visual environment: Anisotropy and the carpentered environment hypothesis. *Vision Research*, 18(10), 1393–1399.

Turrigiano GG (1999). Homeostatic plasticity in neuronal networks: The more things change, the more they stay the same. *Trends in Neurosciences*, 22(5), 221–227.

Ungerleider LG and Mishkin M (1982). Two cortical visual systems. In DJ Ingle, MA Goodale and RJW Mansfield (eds) *Analysis of visual behavior*, pp. 549–586. Cambridge, MA: MIT Press.

Valenza E, Simion F, Cassia VM, and Umiltà C (1996). Face preference at birth. *Journal of Experimental Psychology: Human Perception and Performance*, 22(4), 892–903.

von Melchner L, Pallas SL, and Sur M (2000). Visual behaviour mediated by retinal projections directed to the auditory pathway. *Nature*, 404(6780), 871–876.

Wong ROL (1999). Retinal waves and visual system development. *Annual Review of Neuroscience*, 22, 29–47.

Chapter 3

Learning the best first: interactions between visual development and learning

Melissa Dominguez and Robert A. Jacobs

Introduction

Human infants are born with sensory, cognitive, motor and linguistic limitations relative to adults. For example, infants have blurry vision, limited attentional and memorial capacities and relatively poor control of their eye and hand movements. Counterintuitively, Newport (1988, 1990) hypothesized that mental limitations are useful to the infant (see also Turkewitz and Kenney, 1982). According to her 'Less is more' hypothesis, these limitations simplify the environmental input, thereby making it easier for the child to learn 'rules' that explain the simplified input because the rules themselves are simpler. As the infant matures, mental abilities become less limited and, thus, environmental input becomes less simplified. The simple rules that the infant previously acquired serve as a foundation for the acquisition of more complicated rules based on the more complicated input. Development is useful because it employs a bootstrapping strategy—it is easier to build the complicated rules from the foundation of the early, simplified rules than it is to build them from scratch. This bootstrapping process is continual throughout childhood allowing for the eventual acquisition of knowledge that would have been difficult or impossible to learn with a static, non-developmental system.

Whereas Newport (1988, 1990) studied the less is more hypothesis in the context of language acquisition, we study it with respect to visual development. This chapter considers the hypothesis that visual development aids visual learning by giving the most informative sensory stimuli to the infant first, and gradually adding less informative stimuli to the infant's input as maturation proceeds. We look at this idea in two visual domains: motion velocity estimation and depth perception.

In the case of motion velocity estimation, we consider the importance of the development of visual acuity. Infant acuity is poor relative to that of adults. At birth, a human infant with normal vision has an acuity equivalent to 20/400 as measured on the Snellen scale, compared to 20/20 of a normal adult (Kellman and Arterberry, 1998). A newborn's visual world is effectively low-pass filtered with respect to spatial frequency (i.e. newborn vision is blurry—the newborn senses large visual features but not finer details). Infant acuity improves in a roughly linear fashion over the first eight months of life to near adult levels. During this time period, progressively higher spatial frequency information is added to the child's visual world (i.e. the infant progressively senses finer and finer visual details). This developmental progression is illustrated in Figure 3.1, which shows how a woman's face would look to infants of age one month, two months and three months, and to an adult. We propose that this progression is

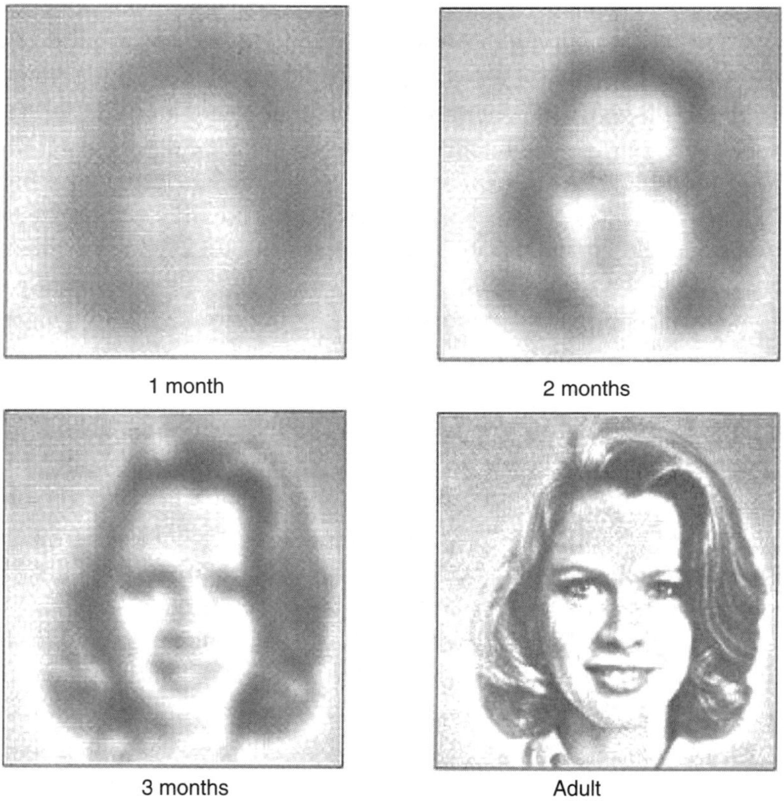

Fig. 3.1 This figure illustrates how a woman's face would look to infants of age one month, two months and three months, and to an adult from a viewing distance of approximately 50 cm. (From Ginsburg, 1983).

helpful for the acquisition of the ability to estimate motion velocities because the low spatial frequency band is the most informative for this task. We provide support for this hypothesis with the results of simulations of artificial neural networks trained to perform this task using image data that is filtered by spatial frequency, and presented to the networks in various sequences. A more extensive presentation of this work can be found in Jacobs and Dominguez (2003).

For the task of depth perception, our approach takes advantage of the fact that there are many cues to visual depth, including motion, stereo and several types of pictorial cues. Here, we concern ourselves only with motion and stereo cues. Infants develop sensitivity to motion before sensitivity to stereo disparity (Kellman and Arterberry, 1998), perhaps because motion is the more reliable information source. In the simulations presented here, we compare the performances of artificial neural networks that use motion and stereo information in various ways. We find that motion is the more informative cue in the given simulation conditions, and also that training with motion alone and then adding stereo information is more effective than the reverse order or than training with the two together from the beginning.

Motion velocity estimation

The first task we study is to visually estimate the velocity of a moving object. Intuitively, one might approach this problem by identifying a visual feature on an object in an image from one moment in time, and then identifying this same feature in an image from a second moment in time. The distance that the feature moved from the first to the second time frame can be used to estimate the object velocity. Under this view of the problem it is easy to see that the low spatial frequency band might contain the most useful information for this task. Low spatial frequencies hold information about large image features such as contours and shapes. High spatial frequencies contain information about small details and edges. Consequently, there are likely to be many false correspondences (i.e. incorrect matches of an image feature from one time frame with an image feature from a second time frame) in the high-frequency domain, and far fewer in the low-frequency domain.

A newborn's visual system effectively filters visual images such that only low spatial frequency information is available to the newborn's perceptual and cognitive processes, with higher spatial frequency information being successively added as development progresses. Previous computational research in the domains of stereopsis (Dominguez and Jacobs, 2003) and object categorization (French, Mermillod, Quinn, Chauvin, and Mareschal, 2002) suggests that this developmental progression is useful for aspects of visual learning. In this section we show that it is also useful for motion velocity estimation, and that the first information available to the infant, low spatial frequency information, is the most reliable for this task. We compare the performances of several models on a motion velocity estimation task. Some models are developmental in that the nature of their input changes during the course of training, whereas others are non-developmental.

Developmental and non-developmental models

We simulated three different kinds of developmental models. The coarse-to-multiscale (C2M) model is based on infant visual development. Early in training, the model received only coarse-scale or low spatial frequency information. As training progressed, information from higher spatial frequency scales was added to its input. The fine-to-multiscale (F2M) model developed in the opposite order, meaning fine-scale or high spatial frequency information was presented to it first, and information at lower scales was added to its input as training progressed. The random developmental (RD) model also had its input changed over the course of time, with more information added at each stage, but these stages were not organized according to spatial frequency content. Instead, the stages were randomly organized. We simulated several non-developmental (ND) models whose input did not change over the course of training; model ND received all of the inputs at all times, and the other non-developmental models received various subsets.

The structure of the developmental and non-developmental models is illustrated in Figure 3.2. The input to each system was a sequence of 88 retinal images, where each image was a one-dimensional array of 40 pixels. As described below, this sequence depicted an object moving at a constant velocity in front of a stationary background. The retinal array was treated as if it were shaped like a circle in the sense that the leftmost and rightmost pixels were regarded as neighbours. This wrap around of the left and right edges was done to avoid edge artefacts. Although a one-dimensional retina is a simplification, its use is justified by the need to keep the simulation times within reason.

The sequence of retinal images was filtered using motion energy filters (see Jacobs and Dominguez, 2003, for more details). Adelson and Bergen (1985) proposed motion energy filters as a way of modelling the motion sensitivities of simple and complex cells in primary visual cortex. Intuitively, these filters may be understood as follows (a more technical description is provided in Box 3.1). The spatiotemporal receptive field profile of a simple cell can often be accurately modelled by a Gabor function (a sine wave multiplied by a normal distribution) over a domain whose dimensions encode both spatial

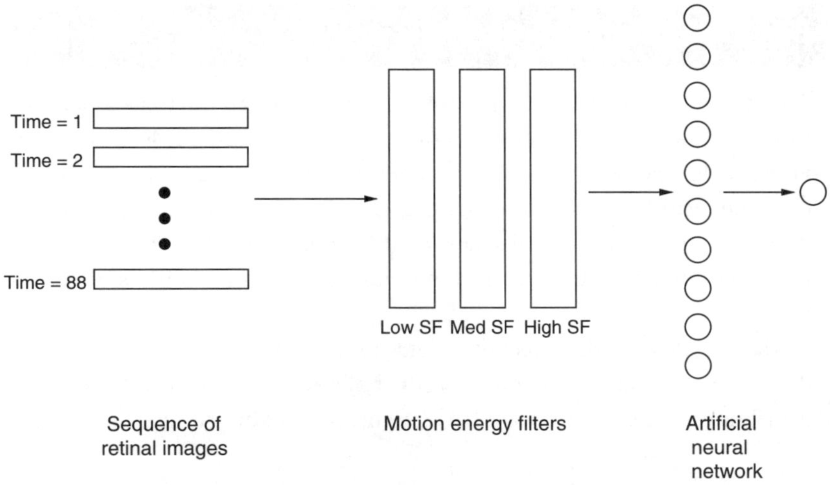

Fig. 3.2 The developmental and non-developmental systems shared a common structure. The input to the systems was a sequence of retinal images. This sequence was filtered by motion energy filters. The set of filters can be partitioned into subsets tuned to low, medium and high spatial frequencies. The outputs of the filters were the inputs to an artificial neural network that was trained to estimate the object velocity depicted in the image sequence.

and temporal parameters. The location, size, and shape of the simple cell's receptive field is specified by the values of the Gabor function's spatial and temporal means, standard deviations, frequencies, and phases.

Consider, for example, the graphs in Figure 3.3. One dimension of each graph represents space, and the other dimension represents time. In the scenarios illustrated in these graphs, there is a one-dimensional object whose length is five pixels. Pixels one, three and five are dark grey, whereas pixels two and four

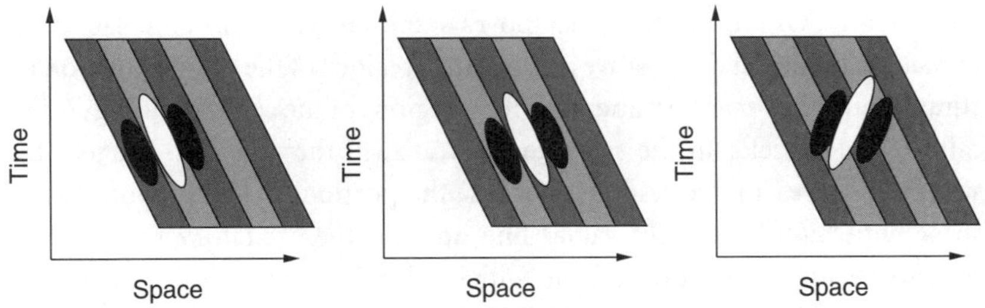

Fig. 3.3 The three graphs illustrate three possible scenarios in which a one-dimensional object (consisting of five pixels: pixels one, three, and five are dark grey, whereas pixels two and four are light grey) moves to the left within the receptive field of a simple cell. See text for explanation.

Box 3.1 Motion energy filters

Adelson and Bergen (1985) proposed motion energy filters as a way of modelling the motion sensitivities of simple and complex cells of primary visual cortex. For ease of explanation, we consider a visual input consisting of a sequence of one-dimensional images. In this case, one can define a two-dimensional spatiotemporal space in which one dimension encodes space in units of pixels, and the other dimension encodes time in units of image frames.

Mathematically, the receptive field profile of a simple cell can be described as a Gabor function (a product of a sine waveform and a normal distribution) tuned to leftward (–) or rightward (+) directions of motion:

$$g^{\pm} = \exp\left\{-\frac{x^2}{2\sigma_x^2} - \frac{t^2}{2\sigma_t^2}\right\} \sin(2\pi[\omega_x x \pm \omega_t t] + \phi) \qquad (1)$$

where x and t are the spatial and temporal distances to the center of the Normal distribution, σ_x^2 and σ_t^2 are the spatial and temporal variances of this distribution, ω_x and ω_t are the spatial and temporal frequencies of the sinusoids, and ϕ is a base phase. A simple cell's preferred velocity is the ratio of its temporal frequency to its spatial frequency. The activity of a simple cell is given by the square of the convolution of the cell's receptive field profile with the spatiotemporal pattern. The activities of two simple cells, one with a base phase of 0 and the other with a base phase of $\pi/2$, are summed in order to form the activity of a complex cell. This method of pooling the responses of simple cells makes the complex cell's response relatively insensitive to where a motion lies within its receptive field.

are light grey. Over time, the object moves to the left. The white and black ovals in each graph depict the positive and negative regions of the Gabor function or, equivalently, the excitatory and inhibitory regions of the corresponding simple cell's receptive field. In the scenario illustrated in the graph on the left, the simple cell gives a large response because the portions of the spatiotemporal space with high luminance values line up with the excitatory region of its receptive field, whereas the portions with low luminance values line up with its inhibitory regions. In contrast, the cell gives small responses in the scenarios illustrated in the middle and right graphs.

In our simulations, the spatiotemporal space is also two-dimensional; one dimension encodes space in pixels and the other dimension encodes time in image frames. The simple and complex cells used a subset of the possible receptive field locations in the two-dimensional (40 pixels $\times$ 88 time frames) spatiotemporal space. This subset formed a 20 $\times$ 4 uniform grid located in the middle of the space such that receptive fields were centered on odd-numbered pixels and odd-numbered time frames. An advantage of this choice of locations was that edge artefacts were avoided because all receptive fields fell entirely within the spatiotemporal space.

Fifteen complex cells corresponding to three spatial frequencies and five temporal frequencies were centred at each receptive field location. All cells were tuned to rightward motion because we restricted our data sets to only include objects that were moving to the right. The spatial and temporal frequencies used by the cells were each separated by an octave. Importantly, temporal frequencies were chosen so that the set of cells at each spatial frequency had the same pattern of velocity tunings. Specifically, the sets tuned to low, medium and high spatial frequencies had velocity tunings of 0.25, 0.5, 1.0, 2.0, and 4.0 pixels per time frame. This was achieved by correlating the spatial and temporal frequency tunings of the cells; i.e. low spatial frequency cells were tuned to a comparatively low range of temporal frequencies, and high spatial frequency cells were tuned to a high range of temporal frequencies. A cell's spatial and temporal standard deviations (i.e. receptive field sizes) were set to be inversely proportional to its spatial and temporal frequencies, respectively. The outputs of the complex cells were normalized so that a cell tended to respond to relative contrast in the image sequence rather than absolute contrast (Heeger, 1992; Nowlan and Sejnowski, 1994).

The outputs of the complex cells were the inputs to an artificial neural network. The network had 1200 input units (the complex cells had 80 receptive field locations and there were 15 cells at each location). The network's hidden layer contained 18 hidden units which were organized into 3 groups of 6 units each. The connectivity of the hidden units was set so that each group had a limited receptive field, and so that neighbouring groups had overlapping receptive fields. A group of hidden units received inputs from 32 receptive field locations at the complex cell level, and the receptive fields of neighbouring groups overlapped by eight receptive field locations. The hidden units used a logistic activation function. The output layer consisted of a single linear unit; this unit's output was an estimate of the object velocity depicted in the sequence of retinal images.

The weights of an artificial neural network were initialized to small random values, and were adjusted during the course of training to minimize a sum of squared error cost function using a conjugate gradient optimization procedure (Press, Teukolsky, Vetterling, and Flannery, 1992). Advantages of this

Table 3.1 A summary of the inputs received by each model at each stage of development

	Model C2M	Model F2M	Model RD	Model ND
Stage 1	Low SF	High SF	Random 1/3	All inputs
Stage 2	Low and medium SF	Medium and high SF	Random 2/3	All inputs
Stage 3	All inputs	All inputs	All inputs	All inputs

procedure are that it tends to converge quickly and it has no free parameters (e.g., no learning rate or momentum parameters). Weight sharing was implemented at the hidden unit level so that corresponding units within each group of hidden units had the same incoming and outgoing weight values, and so that a hidden unit had the same set of weight values from each receptive field location at the complex cell level. This provided the network with a degree of translation invariance, and also dramatically decreased the number of modifiable weight values in the network, thus decreasing the number of data items and the amount of time needed to train the network.

Models were trained and tested using separate sets of training and test data items. Each set contained 250 randomly generated items. Training was terminated after 100 iterations through the training set. The results reported below are based on the data items from the test set.

Three developmental systems and several non-developmental systems were simulated (see Table 3.1). The coarse-to-multiscale system, or model C2M, was trained using a coarse-to-multiscale developmental sequence which was implemented as follows. The training period was divided into three stages. During the first stage, the neural network portion of the model only received the outputs of complex cells tuned to the low spatial frequency (the outputs of other complex cells were set to zero). During the second stage, the network received the outputs of complex cells tuned to low and medium spatial frequencies; it received the outputs of all complex cells during the third stage. The training of the fine-to-multiscale system, or model F2M, was identical to that of model C2M except that its training used a fine-to-multiscale developmental sequence. During the first stage of training, its network received the outputs of complex cells tuned to the high spatial frequency. This network received the outputs of complex cells tuned to high and medium spatial frequencies during the second stage, and received the outputs of all complex cells during the third stage. The training of the random developmental system, or model RD, also used a developmental sequence, though this sequence was generated randomly and, thus, was not based on the spatial frequency tunings of the complex cells. The collection of complex cells was randomly partitioned into three equal-sized subsets with the constraint that each subset included one-third of the cells at each receptive field location. During the first stage of training, the neural network portion of the model only received the outputs of the complex cells in the first subset. It received the outputs of the cells in the first and second subsets during the second stage of training, and received the outputs of all complex cells during the third stage. In contrast, the training period of the non-developmental system, model ND, was not divided into separate stages; its neural network received the outputs of all complex cells throughout the entire training period.

Data sets and simulation results

The performances of the four models were evaluated on two data sets. In all cases the images were grey-scale with luminance values between 0 and 1, and motion velocities were rightward with magnitudes between 0 and 4 pixels per time frame. Fifteen simulations of each model on each data set were conducted.

In the solid object data set, images consisted of a moving light or dark object in front of a stationary grey background. The object's grey-scale values were randomly chosen to either be in the range from 0.0 to 0.1 or from 0.9 and 1.0, whereas the grey-scale value of the background was always 0.5. The size of the object was randomly chosen to be an integer between 6 and 12 pixels, its initial location was a randomly chosen pixel on the retina, and its velocity was randomly chosen to be a real value between 0 and 4 pixels per time frame. Because the velocity was real-valued, the boundaries between an object's internal grey-scale values, or the boundaries between the ends of an object and the background, could fall at a real-valued location within a retinal pixel. In these (common) cases, the luminance value of a pixel was appropriately linearly interpolated. Given a sequence of images, the task of a model was to estimate the object's velocity. The top panel of Figure 3.4 gives an example of ten frames of an image sequence from the solid object data set.

The bar graph in Figure 3.5 illustrates the results. The horizontal axis of each graph gives the model, and the vertical axis gives the root mean squared error (RMSE) on the data items from the test set at the end of training (the error bars give the standard errors of the means). The labels for the developmental models C2M, F2M, and RD include a number. Recall that the training of these models was divided into three training stages (or developmental stages). The number in the label gives the length of developmental stages 1 and 2 (the length of developmental stage 3 can be calculated using the fact that the entire training period lasted 100 iterations). For example, the label 'C2M-5' corresponds to a

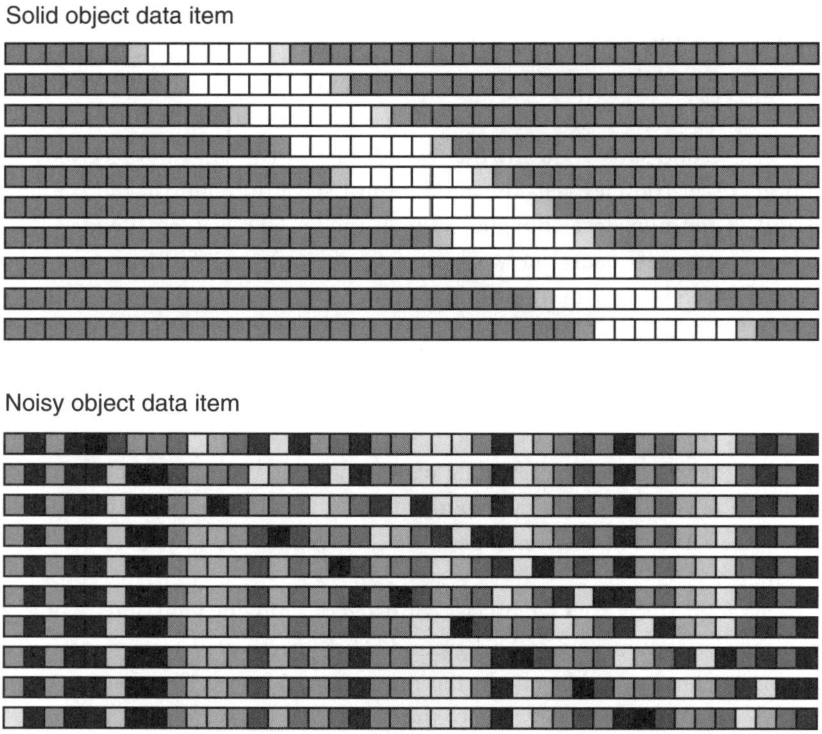

Fig. 3.4 Ten frames of an image sequence from the solid object data set (top) and ten frames of an image sequence from the noisy object data set (bottom). The object in the top panel is moving to the right at a rate of approximately 2.5 pixels per time frame, whereas the object in the bottom panel is moving to the right at a rate of approximately 3 pixels per time frame.

version of model C2M in which the first stage was five iterations, the second stage was five iterations, and the third stage was 90 iterations. In regard to model RD, we simulated four versions of this model (RD-5, RD-10, RD-20, and RD-30). For the sake of brevity, only the version that performed best is included in the graph.

Model C2M significantly outperformed all other models. The version of this model which performed best was version C2M-20 which had an 11.5 per cent smaller generalization error than model ND (the difference between the mean performances of these models is statistically significant; $t(28) = 2.50$, $p < 0.02$ using a two-tailed t-test). In addition, C2M-20 had a 9.6 per cent smaller error than the best version of model F2M ($t(28) = 3.57$, $p < 0.01$), and a 7.2 per cent smaller error than the best version of model RD ($t(28) = 2.30$, $p < 0.05$).

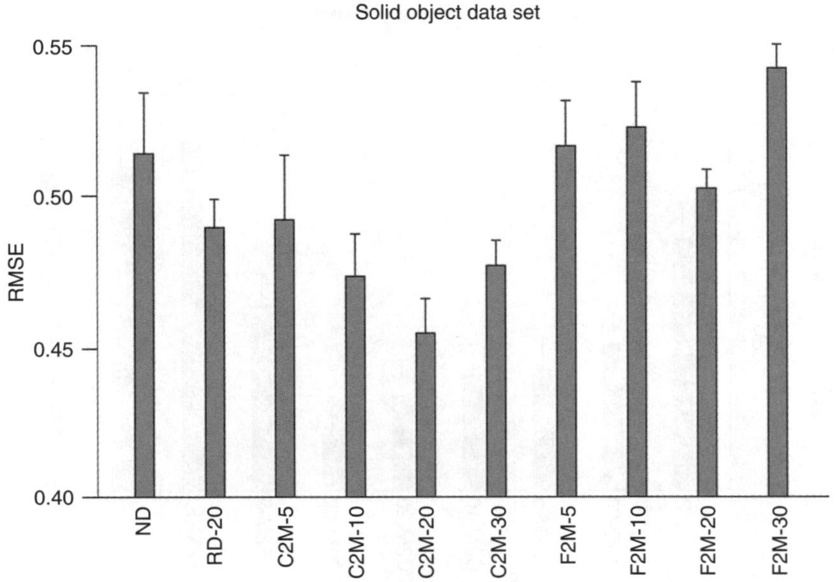

Fig. 3.5 The root mean squared errors (RMSE) on the test set data items for model ND, the best performing version of model RD, and different versions of models C2M and F2M after training on the solid object data set (the error bars give the standard errors of the means).

The images in the second data set, referred to as the noisy object data set, were meant to resemble random-dot kinematograms frequently used in behavioural experiments. Images contained a noisy object which was moving to the right and a noisy background which was stationary. The grey-scale values of the object pixels and the background pixels were set to random numbers between 0 and 1. The size of the object was randomly chosen to be an integer between 6 and 12 pixels, its initial location was a randomly chosen pixel on the retina, and its velocity was randomly chosen to be an integer between 0 and 4 pixels per time frame. As before, the task was to map an image sequence to an estimate of an object velocity. The bottom panel of Figure 3.4 gives an example of ten frames of an image sequence from the noisy object data set.

The results are shown in Figure 3.6. Model C2M, once again, outperformed the other models. Relative to model ND, all versions of model C2M showed superior performance (ND vs. C2M-5: $t(28) = 2.69$, $p < 0.02$; ND vs. C2M-10: $t(28) = 2.78$, $p < 0.01$; ND vs. C2M-20: $t(28) = 3.03$, $p < 0.01$; ND vs. C2M-30: $t(28) = 4.14$, $p < 0.001$). The version of model C2M which performed best was version C2M-30. On average, this version had an 8.9 per cent smaller

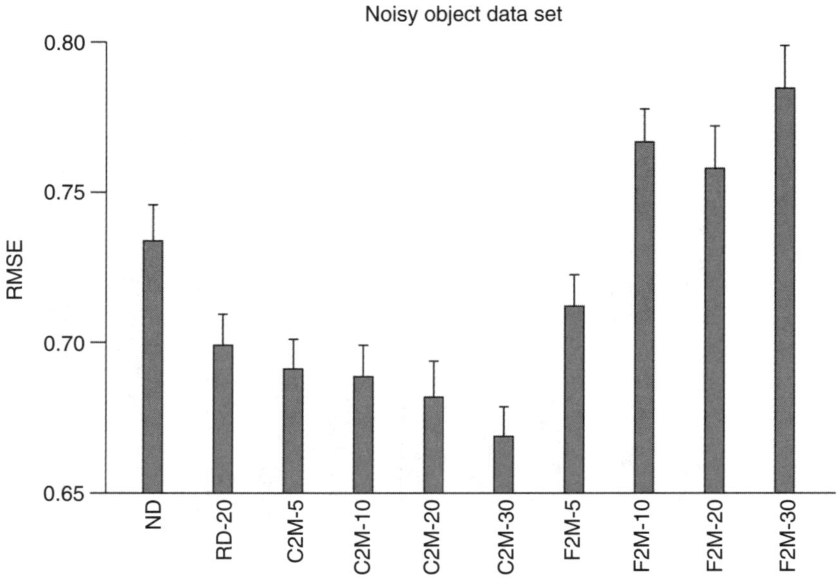

Fig. 3.6 The root mean squared errors (RMSE) on the test set data items for model ND, the best performing version of model RD, and different versions of models C2M and F2M after training on the noisy object data set (the error bars give the standard error of the mean).

generalization error than model ND (t(28) = 4.14, $p < 0.001$), a 6.1 per cent smaller error than the best version of model F2M (t(28) = 2.95, $p < 0.01$), and a 4.3 per cent smaller error than the best version of model RD (t(28) = 2.10, $p < 0.05$).

Taken as a whole, the simulation results are interesting for a number of reasons. Most importantly for our purposes, the fact that model C2M outperformed model ND is important because this demonstrates that a model that undergoes a developmental maturation can acquire a more advanced perceptual ability than one that does not. The fact that model F2M performed similarly to or worse than model ND, and worse than model C2M, is important because this demonstrates that not all developmental sequences provide performance benefits. It is tempting to hypothesize that only those sequences whose characteristics are matched to the task should lead to superior performance. However, the finding that the best version of model RD outperformed model ND is inconsistent with this hypothesis because it is difficult to understand why a random developmental progression would be well-matched to a velocity estimation task. Future work will need to examine this unexpected finding.

To understand these results better, we conducted additional simulations. We compared the performances of several different types of neural networks on the solid object and noisy object data sets. These networks were not developmental systems; each received the same set of inputs throughout the entire training period. Some networks received only the outputs of the low spatial frequency complex cells or motion energy filters (labelled 'coarse' networks), only the outputs of medium frequency filters (labelled 'medium' networks), or only the outputs of high frequency filters (labelled 'fine' networks). Other networks received the outputs of multiple subsets of filters; for example, 'C,M' networks received the outputs of low and medium frequency filters and 'C,M,F' networks received the outputs of all filters (identical to model ND). Lastly, the inputs to 'random' networks were the outputs of a randomly selected one-third of the motion filters. The results at the end of training on the solid object data set are shown in Figure 3.7 (the results on the noisy object data set are qualitatively similar; for the sake of brevity, they are not shown here).

For our purposes, the most important result to emerge from the data is that motion features at different scales are not equally informative for the task of velocity estimation. On the contrary, coarse motion features (i.e. the outputs of low spatial frequency motion energy filters) are more informative than medium-scale motion features which, in turn, are more informative than fine motion features. Evidence for this includes the facts that the coarse network outperformed the medium network, and the medium network outperformed

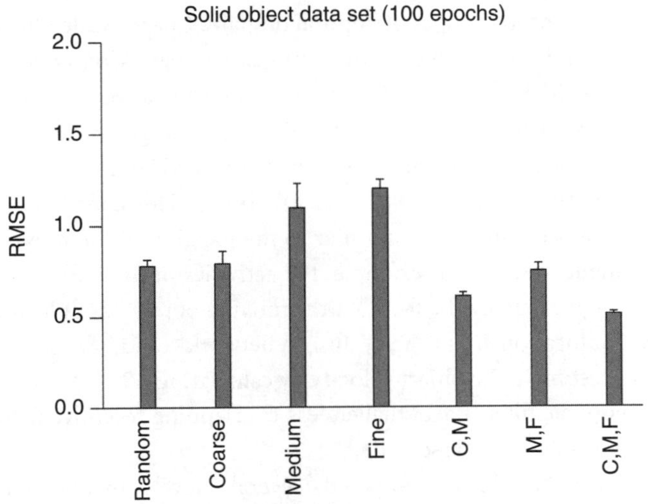

Fig. 3.7 The root mean squared errors (RMSE) of several networks at the end of training on the solid object data set.

the fine network. In addition, the C,M network performed better than the M,F network. These performance rankings were found using both solid object and noisy object data sets.

Based on these results, we can speculate as to why model C2M showed the best performance and why model F2M showed comparatively poor performance. Because the only inputs to model C2M at the start of training were the outputs of the coarse motion energy filters, and because these outputs were the only set that it received at all stages of training, it is likely that this model made extensive use of these signals. The analyses above suggest that the coarse motion signals are the most informative for the velocity estimation task. Model C2M's extensive use of these highly informative signals presumably allowed it to achieve a high level of performance. In contrast, model F2M is likely to have made greater use of the outputs of fine motion energy filters because these outputs were its only inputs at the start of training, and because these outputs were the only set that it received at all stages of training. The analyses above suggest that the fine motion signals are the least informative for the velocity estimation task. Model F2M's extensive use of these least informative signals presumably is responsible for this model's comparatively poor performance.

We have presented analyses suggesting that coarser-scale motion features are more informative than finer-scale features on the velocity estimation task. Several factors may help explain this finding. As discussed by Weiss and Adelson (1998), motion signals tend to be less ambiguous when the stimulus is viewed for a long duration and more ambiguous when the stimulus is viewed for a short duration. Their reasoning applies to the activities of motion energy filters with receptive fields in the spatiotemporal domain. Coarse-scale filters tend to have larger receptive fields than those of fine-scale filters. Consequently, there is less ambiguity in the activities of coarse-scale filters relative to the activities of fine-scale filters. In addition, coarse-scale filters have large, overlapping receptive fields and, thus, form a high resolution coarse-code of the spatiotemporal space (Milner, 1974; Hinton, 1981; Ballard, 1986). This code could provide a network with accurate information as to the location of the moving object at each moment in time. For example, the activities of these filters may have coded with high accuracy the fact that the moving object was at location A at time t_A and at location B at time t_B. If so, a network could have easily learned to accurately estimate the object velocity by calculating $(B - A)/(t_B - t_A)$. In contrast, fine-scale filters have smaller, less-overlapping receptive fields which form a lower resolution coarse-code.[1]

In summary, we have compared several developmental and non-developmental models on a visual motion velocity estimation task. Overall, model C2M performed best, and model F2M often performed worst. The fact that model C2M outperformed model ND is important because this demonstrates that a model that undergoes a developmental maturation can acquire a more advanced perceptual ability than one that does not. The fact that model F2M performed similarly to or worse than model ND, and worse than model C2M, is important because this demonstrates that not all developmental sequences provide performance benefits.

Model C2M's superior performance is interesting because its developmental progression resembles that of human infants. Our simulation results suggest that this developmental sequence may provide important functional benefits for the acquisition of motion velocity estimation. Consistent with the less-is-more hypothesis (Newport, 1988, 1990), our view of perceptual development is that early developmental periods 'set the stage' for later periods in the sense that early periods bias the performance of an organism during these later periods by filtering sensory input such that the most reliable information is presented first.

The stereo-motion system

Stereo disparities (the lateral displacements between image features in the left-eye and right-eye images that arise from the same event in the visual world) and motion disparities (the lateral displacements between features in successive images that arise from the same event in the visual world) are both cues to visual depth. In this section we investigate ways in which these cues can be used by an observer attempting to learn to perceive visual depth. The work is motivated by the fact that in the developing human infant, sensitivities to motion disparities appear before sensitivities to stereo disparities. We hypothesize that this developmental progression is sensible from an information processing perspective because the motion cue is more informative than the stereo cue. In the simulations presented here: (i) motion information provided for more accurate depth classifications than stereo information; (ii) a modular system, in which the two cues were treated separately, outperformed a monolithic system, in which the two cues were treated as an indivisible whole; and (iii) training first with just motion information and then adding stereo information resulted in better performance than any other system that used both information sources. These results are consistent with the above hypothesis.

Data sets

Visual scenes consisted of a background surface and an object. Each image of a scene was a one-dimensional array whose length was 60 pixels. The background was grey (a luminance value of 0.5), and the object was either light or dark (luminance values in the range [0.0, 0.1] or [0.9, 1.0], respectively; image pixels that depicted both the background and the object had their luminance values appropriately interpolated). The distance from the observer to the background was set to 20 pixels, and the observer always fixated the centre of the background. The size of the object was randomly chosen to be a real value between 10 and 25 pixels. Its depth was chosen so that it was closer to the observer than the background by a randomly assigned integer. Its position along the width-axis was a randomly chosen real-valued location.

Several systems were trained to classify the relative depth between the background surface and the object in each visual scene. Two depth classification tasks were used. In the task with two output categories, the object was closer to the observer than the background by either 0, 1, 2, or 3 pixels. Depths of 0 and 1 pixel were classified as category 1, and depths of 2 and 3 pixels were classified as category 2. In the task with five output categories, relative depths were an integer between 0 and 4 pixels. Scenes with identical depth values belonged to the same category.

The simulation results reported below are based on 10 data sets where each set consisted of 500 data items. Whereas some data items were labelled with the correct classification, other items were unlabelled. In general, a 'teacher' does not provide an infant with depth category information—for example, no one tells an infant that the stereo or motion disparities that the infant is currently perceiving corresponds to a surface that is relatively near. So how does the infant learn to perceive visual depth? Berkeley (1709/1910) speculated that visual depth information is obtained at many moments in time by correlating visual sensations with sensations arising from motor movements, such as when simultaneously viewing and grasping an object. If so, then depth category labels are sometimes provided to the visual system based on motor-based sensations. In our simulations, either 10 or 20 per cent of the data items used to train systems were labelled with the correct classification and the remaining items were unlabelled. A system's performance was quantified by the percentage of unlabelled items which the system correctly classified.

A motion cue to the depth relations in a scene was obtained when an observer viewed a scene from 30 viewpoints where successive viewpoints were spaced 6 pixels apart. This sequence of images is an approximation to what the observer would see if he moved his head along a line parallel to the background and object. The velocity of the object is the average difference in the locations of the object in images at successive moments in time. This velocity is a motion parallax cue to the relative depth between the background surface and the object. Sequences of images were filtered using motion energy filters in order to simulate the responses of simple and complex cells in primary visual cortex.

A stereo cue to the depth relations in a scene was obtained when an observer viewed the scene from two viewpoints (referred to as the left-eye and right-eye viewpoints) separated by 18 pixels. The object disparity is the difference in the locations of the object in the left-eye and right-eye images, and this disparity is geometrically related to the relative depth between the background surface and the object. In order to model the binocular sensitivities of simple and complex cells, stereo pairs of images were filtered using a type of filter known as a binocular energy filter (Ohzawa, DeAngelis, and Freeman, 1990). These filters are mathematically similar to motion energy filters. However, they are not sensitive to motion; instead, they are sensitive to shifts in visual features between local patches of left-eye and right-eye images from a single moment in time.

Computational systems and simulation results

The first set of simulations compared a stereo model with a motion model. Each data item for the stereo model consisted of a set of binocular energy filter responses computed on the basis of a visual scene. If the item was labelled, then the item also included the class label for the scene. The labelled and unlabelled data items for the motion model were identical except that they used the motion energy filter responses. This set of simulations allowed us to evaluate whether the stereo or motion information was more useful for the purpose of classifying scene depths.

To perform their classifications, the models used a modification of co-training, a technique developed by Blum and Mitchell (1998) for learning from labelled and unlabelled data. We refer to this modification as iterative *k*-nearest-neighbour (IKNN). In short, unlabelled data items were assigned class labels based on their similarities to labelled items. This process was repeated until all items were assigned a label. Further details can be found in Box 3.2.

Box 3.2 Iterative *k*-nearest neighbour

Iterative *k*-nearest-neighbour (IKNN) is a modification of co-training, a technique developed by Blum and Mitchell (1998) for learning from labelled and unlabelled data items. In our simulations, IKNN was used as follows. At the start of training, the labelled items were placed on a Labelled List and the unlabelled items were placed on an Unlabelled List. At each iteration of the procedure, each item on the Unlabelled List was temporarily assigned a class label and a confidence measure. The temporary

Box 3.2 *(continued)*

label was found by identifying an item's k nearest neighbours from the Labelled List and then allowing those neighbours to vote on a class label (each neighbour had one vote). The confidence measure was the number of neighbours that voted for the temporary class label. If, for example, 7 of 10 of an item's neighbours were labelled as belonging to class 1, then the item would temporarily be assigned a class label of 1 with a confidence measure of 7. Next, for each possible output class, the single item with the highest confidence measure (out of all items temporarily assigned to that class) was moved from the Unlabelled List to the Labelled List and assigned the corresponding class label. For example, if an item temporarily labelled as belonging to class 1 had a confidence measure of seven and all other items with the same temporary class label had a confidence measure of six or less, then the item would be moved from the Unlabelled List to the Labelled List and assigned a class label of 1. (If two or more items had equally high confidence measures, one of the items was selected at random.) This concluded the current iteration, and was followed by the start of the next iteration. Iterations continued until all items were moved from the Unlabelled List to the Labelled List.

k-nearest-neighbour procedures use a similarity metric to identify the nearest neighbours of a data item. Our simulations used a learned similarity metric based on a technique due to Tipping (1999). Consider data modelled by a multivariate normal distribution. A common similarity metric in this case is the Mahalanobis distance which makes use of the covariance matrix of the distribution (Bishop, 1995). Now consider data modeled by a mixture of normal distributions. Tipping studied a generalization of the Mahalanobis metric which makes use of a weighted average of the covariance matrices belonging to the distributions comprising the mixture model (see Tipping, 1999, for more details). An advantage of this similarity metric is that it is sensitive to the clustering properties of the data (e.g., the data may have multiple modes, or the data may have different variances or covariances in different regions). In our simulations, we fit a mixture of five normal distributions with diagonal covariance matrices to the labelled and unlabelled data items for the stereo model (the responses of the binocular energy filters) and again to the data items for the motion model (the responses of the motion energy filters). Using Tipping's (1999) technique, stereo and motion similarity metrics were derived, and these metrics were used by the I*K*NN procedure.

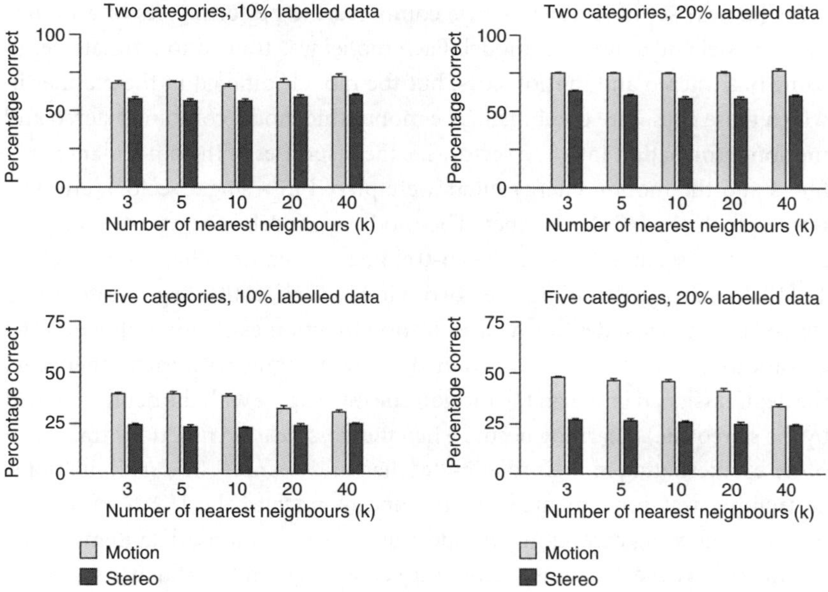

Fig. 3.8 Performance results for the Stereo (dark bars) and Motion Models (light bars). The top two graphs show the results on the classification task with two output categories (the left and right graphs show the results when 10 per cent or 20 per cent of the data items were labelled, respectively), whereas the bottom two graphs show the results on the task with five categories. In each graph, the horizontal axis indicates the number of nearest neighbours that were used when training a system. The vertical axis indicates the average percent of unlabelled data items that were classified correctly by a system (error bars give the standard errors of the means).

The results are shown in Figure 3.8. The results on the categorization task with two depth categories are shown in the top row, and the results on the task with five categories are shown in the bottom row. The left and right columns show the results on simulations with 10 and 20 per cent labelled data items, respectively. The horizontal axis of each graph indicates the number of nearest neighbours used by the IKNN procedure, and the vertical axis indicates the percent of unlabelled data items that a system correctly classified. The light bars are for the motion model, and the dark bars are for the stereo model. On the task with two categories, the motion model shows superior performance in nearly every case. It shows superior performance in every case on the task with five categories. We can conclude, therefore, that motion is the more useful cue for classifying depths under the circumstances considered here. This is not surprising given the fact that the motion information is based on 30 views of a scene, whereas the stereo information is based on two views.

In the next set of simulations we compared the performances of a monolithic model and a modular model. Each model was trained to estimate depth using both stereo and motion cues, but the models differed in the manner in which these cues were combined. The monolithic model combined stereo and motion information into one vector (i.e. the responses of the binocular energy filters and the motion energy filters were placed in a single vector), and was trained with the I*K*NN classifier. The modular model, in contrast, processed the two cues separately using the co-training procedure (Blum and Mitchell, 1998). In short, co-training uses two 'views' of an entity (e.g., motion and stereo cues to visual depth). A classification function exists for each view. The classification functions are constrained to always agree with each other (e.g., the depth assigned based on the motion cue must agree with the depth assigned by the stereo cue). Learning occurs when the classification functions train each other by alternating in their roles as 'teacher' and 'learner'. In our simulations, co-training was implemented in the same manner as the I*K*NN procedure with the following exception. On odd-numbered iterations, data items in the Unlabelled List were assigned temporary class labels and confidence measures on the basis of the motion information. These values were assigned on the basis of stereo information on the even-numbered iterations. That is, the use of motion and stereo information alternated. In this way, depth-from-motion judgements were used as 'teaching' signals for the estimation of depth-from-stereo judgements, and vice versa.

The results are shown in Figure 3.9. The Modular Model outperformed the Monolithic Model in all cases. Under the conditions studied here it is better to treat the two cues separately when making depth classifications than it is to mix them.

Our final set of simulations compared two models that learned from visual cues in a two-stage process. In the first stage, the model classified one-half of the unlabelled data items on the basis of only one cue, using the I*K*NN classifier. In the second stage it classified the remaining data items using co-training as in the modular model above. The M2Mod model used motion data in the first stage, and the S2Mod model used stereo data in the first stage. The results can be seen in Figure 3.10. M2Mod outperformed S2Mod in every case. In fact, a comparison of the graphs in Figures 9 and 10 shows that M2Mod performed best out of all the models that used both stereo and motion information.

The simulations reported in this section were motivated by the fact that infants develop sensitivities to motion disparities before they develop sensitivities to stereo disparities. We hypothesized that this developmental progression is sensible from an information processing perspective because the motion

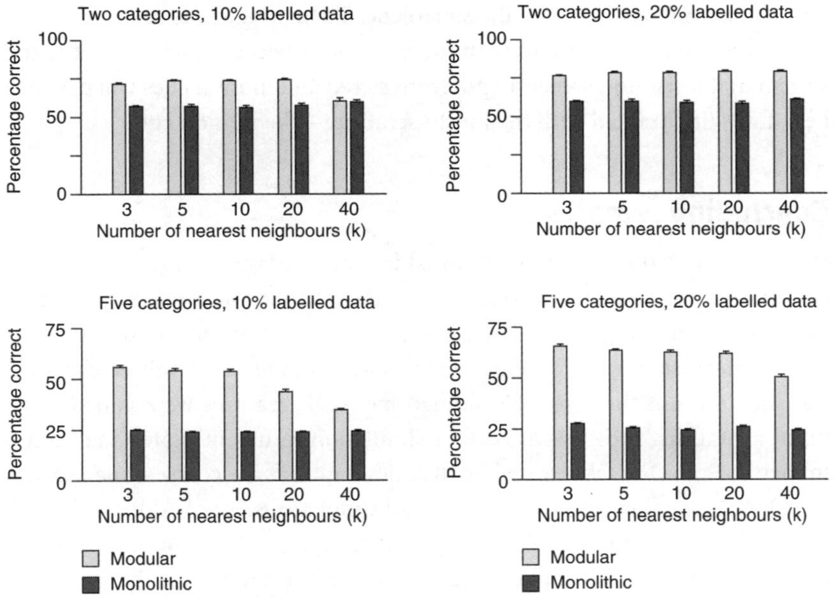

Fig. 3.9 Performance results for the modular and monolithic models.

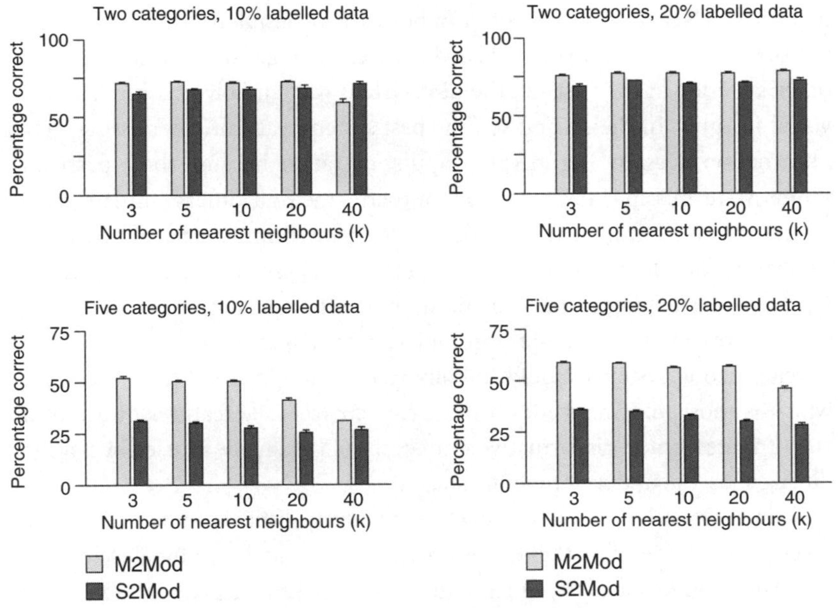

Fig. 3.10 Performance results for the M2Mod and S2Mod models.

cue is more informative than the stereo cue. Taken as a whole, the simulation results are broadly consistent with the hypothesis because they suggest that a system that learns to perceive depth from stereo and motion cues will perform best if it is first trained with the most useful cue—the motion cue.

Concluding remarks

We described two sets of simulations addressing the issue of how visual development interacts with visual learning. In the first set, we considered the task of visually estimating the velocity of a moving object. We found that the best performing system was a developmental system that received only coarse-scale visual features at the start of training; finer-scale features were added to its input as training progressed. Further simulation results indicated that useful information for the task existed at all scales, but that the coarse-scale features were the most reliable. In the second set of simulations, we considered the task of visually estimating the relative depth of an object from motion and stereo cues. We found that the motion cue was more informative than the stereo cue for the task, and that a system that was initially trained solely with the motion cue before being trained with both cues outperformed all other systems that were trained with both cues.

Overall, we believe that our results are consistent with the less-is-more hypothesis (Newport, 1988, 1990). In both sets of simulations the system that performed best was missing inputs during early stages of learning relative to other systems. In the first set, the best system was initially missing fine-scale visual features. In the second set, the best system was initially missing stereo information. Despite the missing inputs, or rather because these particular inputs were missing, these systems outperformed all others. Initially, these systems received simplified views of the visual environment that allowed them to acquire simple knowledge representations. These simple representations were useful 'stepping stones' for the subsequent acquisition of more complicated representations. In addition, our results suggest that the inputs which developmental systems should initially receive should come from the source which is most informative for a task (e.g., coarse-scale features or a motion cue). Inputs from less informative sources should gradually be used by a system during subsequent developmental stages.

From a neurobiological perspective, relatively little is known about the mechanisms underlying the development of visual acuity or the development of sensitivities to visual depth cues. In regard to visual acuity, many aspects of the eye and the optical and retinal elements involved in visual perception are initially immature, and improve postnatally (see Kiorpes and Movshon, 2004, for details). For example, the eye grows from infancy to adulthood thereby increasing the magnification of the retinal image, the quality of the eye's optics improves thereby increasing the resolution and contrast of the retinal image, and the morphology and distribution of cone photoreceptors undergo marked changes after birth. However, optical and retinal changes are thought to play only a minor role, and the major factors contributing to increased spatial resolution and sensitivity to spatial contrast are currently unknown or poorly understood neural factors (Banks and Bennett, 1988; Kiorpes and Movshon, 2004). In regard to the development of sensitivities to visual depth cues, this topic has received very little attention from neuroscientists. Future neuroscientific work will, hopefully, address these important issues.

Acknowledgements

This work was supported by NSF Graduate Fellowship DGE9616170 and by NIH research grant R01-EY13149. We thank R. Aslin for many interesting discussions.

Glossary

Binocular energy filter This filter extracts binocular information from a pair of left-eye and right-eye images (such as a shift of an image feature between the two images). It consists of two stages. The first stage uses left-eye and right-eye Gabor filters in order to extract binocular shift information in a local image region. The second stage combines the outputs of multiple Gabor filters in order to extract binocular shift information in a local region in a way that is relatively invariant to where the shift occurs within the region. The first and second stages are often intended to mimic the receptive field characteristics of simple and complex cells of primary visual cortex, respectively.

Coarse coding A way of forming a distributed representation of the value of a variable that is often used with artificial neural networks. The resulting representation resembles the neural representation of the value of a sensory variable formed by topographic maps in primary sensory areas of primate cortex.

Conjugate gradient A procedure for finding the values of a set of parameters that optimize a function. This procedure can be used, for example, to find the values of the weights of an artificial neural network that minimize a cost function.

Cost function A mathematical function that specifies how the designer of a system wants the system to perform. For example, artificial neural networks are often trained to optimize a cost function which states that the network's output should match the target output as closely as possible.

Gabor filter A smoothly tapered segment of a sinusoid obtained by multiplying a sinusoid with a Normal distribution. The filter has a well-defined mean, standard deviation, frequency and phase. It is often used to mimic the receptive field characteristics of simple cells in primary visual cortex.

Iterative *k*-nearest-neighbour A procedure for classifying a data item based on the class labels assigned to similar data items (see Box 3.2).

Logistic activation function Each unit of an artificial neural network maps its weighted sum of inputs to an output value, referred to as an activation. The logistic activation function maps a weighted sum of inputs to a real-valued activation between 0 and 1.

Mahalanobis distance A metric for measuring the distance between two vectors that takes into account the variances of vector elements and the covariances among pairs of vector elements.

Motion energy filter This filter extracts visual motion information from a sequence of images. It consists of two stages. The first stage uses Gabor filters defined over a spatiotemporal space (e.g., a temporal sequence of visual images) in order to extract visual motion information in a local region of the space. The second stage combines the outputs of multiple Gabor filters in order to extract motion information in a local region of a spatiotemporal space in a way that is relatively invariant to where the motion occurs within the region. The first and second stages are often intended to mimic the receptive field characteristics of simple and complex cells of primary visual cortex, respectively.

Multivariate normal distribution This is a type of probability distribution for a random variable that is a vector (i.e. the variable contains multiple elements such as the height and weight of a person, or the temperature, humidity, and barometric pressure at a spatial location).

Spatial frequency Fourier series approximate the values of a function by a sum of sinusoids of different frequencies, amplitudes, and phases. When the function is parameterized by a spatial variable, such as a function that gives the grey-scale value of each pixel in an image, the frequency of a sinusoid is referred to as a spatial frequency.

Stereopsis The process by which we perceive a three-dimensional visual environment based on two-dimensional left-eye and right-eye retinal images.

Because our left and right eyes slightly offset from each other, left-eye and right-eye images of a scene differ.These differences are known as binocular disparities. They are interesting because they carry information about the three-dimensional structure of a visual environment. The depth information required to perceive the third dimension is computed based on these binocular disparities.

Temporal frequency Fourier series approximate the values of a function by a sum of sinusoids of different frequencies, amplitudes, and phases. When the function is parameterized by a temporal variable, such as a function that gives the grey-scale value of a pixel in each image in a temporal sequence of images, the frequency of a sinusoid is referred to as a temporal frequency.

Weight sharing Refers to the practice of forcing two or more weights of an artificial neural network to have the same values at all times. This practice is done to constrain a neural network to exhibit some desirable property, such as translation invariance.

Note

1. Analyses by Hinton (1981) and Ballard (1986) provide an understanding of the relationship between receptive field size and resolution in the case of binary units that each become active when a stimulus falls within its receptive field. If D is the diameter of a unit's receptive field, k is the dimensionality of the space to be represented, and N is the desired number of just-noticeable differences in each dimension (i.e. the desired resolution), then the required number of units is N^k / D^{k-1}. For a fixed number of units, a high resolution code (that is, one with a large N) requires units with large receptive fields (fields with a large D), whereas a low resolution code can be achieved with units with small receptive fields.

References

Adelson, EH and Bergen, JR (1985). Spatiotemporal energy models for the perception of motion. *Journal of the Optical Society of America A*, 2, 284–299.

Ballard, DH (1986) Cortical connections and parallel processing: Structure and function. *Behavioral and Brain Sciences*, 9, 67–120.

Banks, MS and Bennett, PJ (1988) Optical and photoreceptor immaturities limit the spatial and chromatic vision of human neonates. *Journal of the Optical Society of America A*, 5, 2059–2079.

Berkeley, G (1709/1910) *Essay towards a new theory of vision*. London: Dutton.

Bishop, CM (1995) *Neural networks for pattern recognition*. New York: Oxford University Press.

Blum, A and Mitchell, T (1998) Combining labelled and unlabelled data with co-training. In *Proceedings of the eleventh annual conference on computational learning theory*. New York: ACM Press.

Dominguez, M and Jacobs, RA (2003) Developmental constraints aid the acquisition of binocular disparity sensitivities. *Neural Computation*, 15, 161–182.

French, RM, Mermillod, M, Quinn, PC, Chauvin, A and Mareschal, D (2002) The importance of starting blurry: Simulating improved basic-level category learning in infants due to weak visual acuity. In *Proceedings of the 24th annual conference of the Cognitive Science Society*. Hillsdale, NJ: Lawrence Erlbaum Associates.

Ginsburg, A(1983) *Contrast perception in the human infant*. Unpublished manuscript.

Heeger, DJ (1992) Normalization of cell responses in cat striate cortex. *Visual Neuroscience*, 9, 181–197.

Hinton, GE (1981) Shape representation in parallel systems. In A Drina (ed.) *Proceedings of the Seventh International Joint Conference on Artificial Intelligence*.

Jacobs, RA and Dominguez, M (2003) Visual development and the acquisition of motion velocity sensitivities. *Neural Computation*, 15, 761–781.

Kellman, PJ and Arterberry, ME (1998) *The cradle of knowledge: Development of perception in infancy*. Cambridge, MA: MIT Press.

Kiorpes, L and Movshon, JA (2004) Neural limitations on visual development in primates. In LM Chalupa and JS Werner (eds) *The visual neurosciences*, pp. 159–173. Cambridge, MA: MIT Press.

Milner, PM (1974) A model for visual shape recognition. *Psychological Review*, 81, 521–535.

Newport, EL (1988) Constraints on learning and their role in language acquisition: Studies of the acquisition of American Sign Language. *Language Sciences*, 10, 147–172.

Newport, EL (1990) Maturational constraints on language learning. *Cognitive Science*, 14, 11–28.

Nowlan, SJ and Sejnowski, TJ (1994) Filter selection model for motion segmentation and velocity integration. *Journal of the Optical Society of America A*, 11, 3177–3200.

Ohzawa, I, DeAngelis, GC, and Freeman, RD (1990) Stereoscopic depth discrimination in the visual cortex: Neurons ideally suited as disparity detectors. *Science*, 249, 1037–1041.

Press, WH, Teukolsky, SA, Vetterling, WT, and Flannery, BP (1992) *Numerical recipes in C: The art of scientific computing*. Cambridge: Cambridge University Press.

Tipping, ME (1999) Deriving cluster analytic distance functions from Gaussian mixture models. In *Proceedings of the ninth international conference on artificial neural networks*. Edinburgh.

Turkewitz, G and Kenney, PA (1982) Limitations on input as a basis for neural organization and perceptual development: A preliminary statement. *Developmental Psychobiology*, 15, 357–368.

Weiss, Y and Adelson, EH (1998) *Slow and smooth: A Bayesian theory for the combination of local motion signals in human vision*. Center for Biological and Computational Learning Paper Number 158, Cambridge, MA: Massachusetts Institute of Technology.

Chapter 4

Why let networks grow?

Thomas R. Shultz, Shreesh P. Mysore, and Steven R. Quartz

'Let it grow, Let it grow.'

Grateful Dead, 1973, Eric Clapton, 1974, Boogie Brown Band, 2003

Introduction

Beyond the truism that to develop is to change, the process of developmental change itself has been relatively neglected in developmental psychology. Methodologically, most behavioural studies have traditionally utilized a cross-sectional approach. This has revealed a great deal about how certain behavioural and cognitive abilities differ at various points in development, but it tends to reveal less about the developmental processes that operate to transform those cognitive and behavioural abilities. A variety of factors account for the lack of explanations of developmental change, ranging from the methodological challenges their study entails to principled, learning-theoretic arguments against the very existence of such developmental processes (Macnamara, 1982; Pinker, 1984).

Among the most influential arguments against developmental change was Chomsky's (1980) instantaneity argument that starts with the null hypothesis that children are qualitatively similar learners to adults and supposes that development is instantaneous, in the sense that there are no time dependencies in development. Departing from this view and supposing that children are initially more restricted learners only results in weakening their acquisition properties. The upshot is that *less is less*, and development only reduces learning capability. Thus, as a sort of charity principle, one ought to at least start with the hypothesis that children do not differ substantially from adults, because explaining development is difficult enough without having to do so with greatly diminished learning ability.

These sorts of arguments led to the widespread assumption that there is little theoretical insight to be gained from studying processes of developmental change. Indeed even with the advent of connectionist models and their adoption in developmental psychology, most of the models that were used were qualitatively similar to models of adult learning. That is, most models used a fixed feedforward architecture in which the main free parameters were connection strengths. Therefore such models implicitly adopted Chomsky's argument and began with the assumption that the immature and mature states are qualitatively identical.

Yet how well-founded is this assumption? Although Chomsky's approach is often considered to be a biological one, it is actually quite unbiological in the sense that it is not developmental. Genuine biological approaches invariably emphasize the importance of development (Gilbert, 2003). Viewed from the perspective of developmental cognitive neuroscience, it is now well established that the structural features of neural circuits undergo substantial alterations throughout development (Quartz and Sejnowski, 1997). Do such changes add anything of interest to the explanation of developmental change? If so, does a better understanding of the processes of developmental change undermine Chomsky's argument, and thereby demonstrate that developmental change is crucial for understanding the nature of cognitive development?

In this chapter we ask the question, why let networks grow? We begin by reviewing the wealth of accumulating data from neuroscience that network growth appears to be a much more central feature of learning than is traditionally assumed. Indeed it appears that the assumption that the main modifiable parameters underlying learning are changes in connection strength in an otherwise fixed network may need revision. We discuss this question with an eye toward evidence for learning-directed architectural changes in the brain throughout the lifespan. Next, we consider the computational implications of learning-directed growth. There we consider whether less really is less, or whether by breaking down the traditional distinction between intrinsic maturation and cognitive processes of learning, the learning mechanism underlying cognitive development thereby becomes substantially different, and more powerful, in its learning properties than a fixed architecture. We then present a case in which activity-dependent network growth underlies important acquisition properties in a model system, auditory localization in the barn owl. This non-human system is very well characterized in terms of biological underpinnings, which provide important clues into the likely biological mechanisms underlying human cognitive development, and which give rise to general computational principles of activity-dependent network growth. Then we present evidence from modelling cognitive development in children. There, we explore a number of computational simulations utilizing the cascade-correlation (CC) algorithm. In contrast to the more commonly used back-propagation algorithm, CC starts with a minimal network and adds new units as a function of learning. CC can be viewed as an abstract rule for neural growth, and thus as a means of exploring the computational and learning properties of such developmental processes. In the final section we present the main conclusions from this work and point to areas of future research and open research questions.

Experience-dependent architectural plasticity

Two broad categories of neural plasticity can be distinguished based on the nature of expression and encoding of change in the nervous system. They are 'synaptic efficacy change' and 'architectural plasticity'. The former, as the name suggests, refers to changes to existing, functional synapses in a network whose topology is kept fixed, while the latter refers to changes in the underlying topology of the network. Whereas synaptic efficacy change has been the predominantly studied form of plasticity (Martin and Morris, 2002), neuroscience research also provides evidence for activity-dependent plasticity. Structural mechanisms largely implement architectural plasticity, and they include dendritic spine motility, spinogenesis (the induction of new spines), synaptogenesis (the formation of new synapses), and neurogenesis (the growth of new neurons). We provide a brief sketch of the research literature in these areas.

Most excitatory synapses in the brain are formed on small protrusions from neuronal dendrites called spines. Spines are extremely important postsynaptic structures containing several neurotransmitter receptors, and various biochemical signaling mechanisms, among other elements. Based on the synaptic input they receive, spines trigger different signaling pathways that can result in short-term or long-term synaptic changes. Rapid morphological plasticity of spines, spine movement and their growth and retraction are collectively referred to as spine motility. Motility was found to occur spontaneously in hippocampal neurons in slice cultures from rats (Dailey and Smith, 1996; Fischer *et al.*, 1998). Since its discovery, one of the main functions attributed to motility is that of exploring the extracellular space in search of presynaptic partners. This view is supported by the finding that formation of synapses stabilizes previously mobile dendritic protrusions (Dailey and Smith, 1996). Motility has also been found in response to electrical stimulation of neurons (Maletic-Savatic *et al.*, 1999), thus suggesting that experience may play a role in regulating motility. Direct evidence for this possibility stems from the finding that trimming the whiskers of rats, thereby causing significant sensory deprivation, resulted in a 40 per cent drop in spine motility in corresponding whisker barrels in the rodent brain (Lendvai *et al.*, 2000).

While the above studies looked directly at movement and morphological changes as they occurred, other studies have looked at the effects of motility indirectly by comparing the density of spines and their morphological distribution between control and learning conditions. In a trace eye-blink conditioning task in rats, it was found that the density of spines in the basal CA1 dendrites of the hippocampus—an area that is known to be important for encoding such memories—was greater after conditioning (Leuner *et al.*, 2003). Interestingly, when the acquisition of this association was blocked using a pharmacological agent that interferes with the formation of long-term memories, this spine density increase was blocked as well. Spatial training of rats, which is known to produce an increase in their subsequent ability to learn in spatial tasks, also produces an increase in the density of spines in the hippocampus (Moser *et al.*, 1997). Other tasks like odour discrimination training (Knafo *et al.*, 2001), visual stimulation, and even space flight (Yuste and Bonhoeffer, 2001), have been found to produce spine density increases in the appropriate brain regions in rats. A case can thus be made that experience-dependent plasticity in dendritic spines may facilitate architectural reorganization of neural circuits in response to functional demands.

In addition to the above studies that examined spine changes, other studies have looked directly at changes in synapse numbers in response to learning. For instance, Chang and colleagues (1991) showed that long-term potentiation increased synaptic numbers in two-year-old rats *in vitro*. Other studies report an increase in the cerebellar synapse density in rats following classical eye-blink conditioning experiments (Black *et al.*, 1990; Kleim *et al.*, 2002), and synaptogenesis in the hippocampus following spatial training of rats in the Morris water maze task, a classic spatial learning experiment (Ramirez-Amaya *et al.*, 1999). Such results are not restricted to rodents (for instance, see Stewart and Rusakov, 1995, for similar effects in chicks). Experimental evidence for direct sensory-stimulus-dependent synapse formation in adult animals was first reported in rodent whisker barrels, where it was shown that localized increases in both synapse number (35 per cent) and spine density (25 per cent) followed specific whisker stimulation (Knott *et al.*, 2002; Zito and Svoboda, 2002). Thus, several lines of evidence establish the occurrence of synaptogenesis in adult neuronal tissue following physiologically relevant stimuli and learning paradigms.

Finally, we turn to neurogenesis in adult animals. In traditional neuroscience, the idea that neurons may be added continually after birth was not considered seriously (Gross, 2000). Since the early 1960s, this view gradually began to change when Altman and colleagues (Altman, 1962; Altman and Das, 1965) showed evidence for new neurons in several brain regions of adult rats (neocortex, olfactory bulb, dentate gyrus), and cats (neocortex). Since then, evidence for neurogenesis has accumulated in adult songbirds (Nottebohm, 1985; Goldman and Nottebohm, 1983; Paton and Nottebohm, 1984), dentate gyrus of rats (Stanfield and Trice, 1988), adult mouse hippocampus (Kempermann *et al.*, 1997), and dentate gyrus and olfactory bulb of macaque monkeys (Kornack and Rakic, 1999). Recent studies have even reported the addition of new neurons in the neocortex of macaque monkeys throughout adulthood (Gould *et al.*, 1999b). We note that the results in the neocortex are still controversial as some researchers have expressed doubt about them based on objections to the techniques used to establish neurogenesis (Nowakowski and Hayes, 2000; Rakic, 2002). In humans, it has been found that new neurons are added in the dentate gyrus (Eriksson *et al.*, 1998), but not to the olfactory bulb (Sanai *et al.*, 2004). The latter study points to the presence of a substantial number of adult neural stem cells—cells that can potentially generate neurons—in a region of the forebrain called the subventricular zone, that intriguingly appear not to produce neurons in the adult human brain.

The above evidence indicates that neurons are indeed added to adult mammalian neuronal circuits, although this addition appears to be less prevalent in humans. An important question that remains is whether this neurogenesis is functionally important or merely a 'vestige of development' (Gross, 2000; Kempermann, 2002; Nottebohm, 2002), especially since the functional integration of new neurons into well-developed circuits may be difficult (Rakic, 2004). The following considerations suggest that neurogenesis does, in fact, mediate function. New neurons are added to structures that are important for learning and memory, such as the hippocampus, lateral prefrontal cortex, inferior temporal cortex and posterior parietal cortex. Neurogenesis, like synaptogenesis and spine density changes, has been correlated with hippocampally dependent learning experiences. For instance, trace eye-blink conditioning and spatial learning in animals lead to an increase in the number of neurons through an extension of neuronal lifetimes (Gould *et al.*, 1999a). There appears to be a critical period following cell production such that learning occurring in this period increases neuronal lifespan. In parallel to this evidence, stressful experiences that result in a downregulation of cell proliferation in the dentate gyrus (Gould and Gross, 2002) are implicated in lower performance in hippocampally dependent learning tasks, suggesting a causal link between the two. Several conditions that increase neurogenesis (enriched environments, increased oestrogen levels, wheel running, etc.) in mice and rats also enhance performance (Gould and Gross, 2002). Similarly, increases in social complexity have been found to enhance the survival of new neurons in birds (Lipkind *et al.*, 2002). Finally, it has been found that the physiological properties of adult-generated granule cells in the dentate gyrus of the hippocampus resemble those of granule cells in young rats (Overstreet-Wadiche *et al.*, 2006). This suggests that adult-generated neurons may share some properties with embryonic and early postnatal neurons in their ability to extend axons, form new connections more readily and to make more synapses. These characteristics may make adult neurogenesis an attractive 'feature', rather than a developmental 'bug' in neuronal circuits.

In summary, there is now a large body of evidence that suggests that spine motility, spinogenesis, synaptogenesis, and more recently, neurogenesis are all active mechanisms for implementing architectural plasticity in neuronal circuits in response to real learning and memory needs faced by organisms.

Computational importance of learning-directed growth

Formal learning theory (FLT) deals with the ability of a learner to arrive at a target concept based on examples of the concept. Three typical features of FLT models are that the learning algorithm searches through a predefined space of candidate hypotheses (concepts), it is expected to learn the concept exactly, and no restrictions are placed on the actual time taken by the learner to arrive at the target concept. FLT is therefore interested in 'exact' and 'in-principle' learnability, and the expectation is that generalization—a measure of performance on novel examples—is achieved. The classical formulation of FLT is discussed in the context of language learning by Gold (1967). Perhaps the key insight from formal work on language learning is that the learner must utilize a highly restricted set of all possible concepts in order to have even the possibility of generalizing. In other words, far from employing a general learner, from this perspective a language learning system must be meticulously tailored to the problem at hand, either by the designer in the case of artificial systems, or presumably by evolution in the case of biological learners.

In the early 1980s, formal learning theory underwent a substantial change from Gold's limit-based framework to a model of probably-approximately-correct (PAC) learning (Valiant, 1984). The PAC model relaxes two of the three key requirements of formal learning theory. The impractical assumption of infinite time horizons is eliminated and the stringent restriction of exact learning relaxed. In short, PAC learning deals not with in-principle learnability, but with learnability in polynomial time, and not with exact learning, but with approximate learning. Formally, a concept is PAC-learnable if the learner arrives with probability 1-d at a hypothesis that classifies all the examples correctly with probability 1-e, for arbitrarily small e and d. Nevertheless, the hypothesis space is still fixed a priori. A fixed hypothesis space yields such problematic theoretical issues as Fodor's paradox (Fodor, 1980), which states that nothing can be learned that is not already known; and hence nothing is really learned. The idea here is that no hypothesis that is more complex than the ones in the given hypothesis space can be evaluated and hence learned. Therefore, all concepts need to be available in the hypothesis space before the search begins.

Constructive learning (Piaget, 1970, 1980; Quartz, 1993; Quartz and Sejnowski, 1997) addresses this issue of a fixed hypothesis space. The central idea of Piagetian constructivism is the construction of more complex hypotheses from simpler ones. This issue is dealt with more formally by Quartz (Quartz, 1993; Quartz and Sejnowski, 1997). Constructivist learning models deal directly with the issue of increasing hypothesis complexity as learning progresses (Shultz *et al.*, 1995; Westermann, 2000). Activity-dependent structural plasticity is viewed as the mechanism that implements constructivist learning. Constructive neural networks offer a clear way of viewing learning and development as constituting a 'plasticity continuum'. Synaptic weight change may be a form of plasticity that occurs at fast timescales, whereas architectural changes occur on slower timescales. Further, it is also possible, as we saw earlier, that the developmental processes of structural plasticity underlie learning even in mature animals.

Baum (1988) showed that networks with the ability to add structure during the process of learning are capable of learning in polynomial time any learning problem that can be solved in polynomial time by any algorithm whatsoever (Quartz and Sejnowski, 1997), thus conferring a computational universality to this paradigm. Interestingly, the bias-variance trade-off can be broken by constructivist learning, by adding hypotheses incrementally to the space in such a way as to keep variance low while reducing bias. Further, in the context of neurobiology, the burden of innate knowledge is relaxed. Given a basic set of primitives (in the form of mechanisms and physical substrate), construction of a network occurs under the guidance of experience within genetic constraints. Finally, computational arguments show that it is unlikely that evolution has prepared brain networks in human children for all of the various learning problems to which they might eventually be exposed (Sirois and Shultz, 1999). It is far more likely that brain networks will need to be constructed and their architectures modified as novel and unexpected learning problems arise.

Architectural plasticity and auditory localization in barn owls

The auditory localization system (ALS) of the barn owl is an excellent example of representational change that is produced by experience-driven architectural plasticity. Beacause developmental mechanisms and programmes appear to be highly conserved across many species and phyla, systems like this may reveal important general principles of neurobiological self-organization and mechanisms by which instructive environmental signals play a role in the construction of representational structures that also most likely operate in human development. Whereas many animals can localize sounds soon after birth (Field *et al.*, 1980) indicating the hard-wiring of at least a part of this system (Brainard and Knudsen, 1998; Knudsen *et al.*, 1991), experience, in addition to intrinsic programmes, plays an important role in shaping and modifying the auditory localization pathway in barn owls.

Visual displacement experiments in barn owls

Barn owls localize sound in the horizontal direction primarily using an auditory cue called the interaural time difference (or ITD). This is defined as the delay or the time difference between the arrival of sound from a source at one ear versus the other. Unless the sound source is positioned symmetrically with respect to the two ears, the signal has to travel different distances to reach the near versus the far ear, resulting in a non-zero ITD.

One of the pathways in which ITD information is processed in barn owls is the midbrain pathway (see Figure 4.1). Here, cue information arrives in a frequency-dependent manner at the central nucleus of the inferior colliculus (ICC), and flows through the external nucleus of the inferior colliculus (ICX) to the optic tectum (OT). The ICC has a tonotopic, or frequency-dependent, organization, and cue information from various frequency channels of the ICC is combined (shown as thick arrows) to form a topographic, map-like representation of auditory space in the ICX. The ICX then conveys this topographic information about auditory space to the OT (Knudsen, 2002). Interestingly, the OT has a similar topographic map of *visual* space mediated by visual input from the retina and forebrain. It has been found that the auditory and visual maps of space are aligned and integrated in the OT (Knudsen and Brainard, 1995).

Visual displacement experiments have explored the nature of information processing in the midbrain pathway and have shed light on the mechanisms by which plasticity occurs in response to external demands. These experiments involve the use of prismatic spectacles to disrupt auditory-location

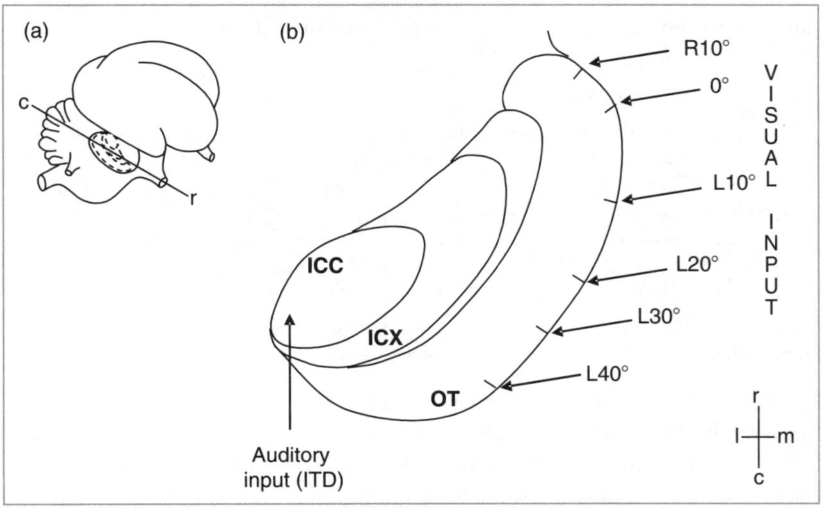

Fig. 4.1 Midbrain auditory pathway in barn owls. (a) Cartoon showing lateral view of a barn owl's brain. The straight line represents the plane of section through the midbrain. (b) Schematic representation of the sectioned surface of the *right* midbrain. Shown are the three important processing areas in the midbrain localization pathway: ICC—central nucleus of the inferior colliculus, ICX—external nucleus of the inferior colliculus, and OT—optic tectum. The arrows on the left and the right show respectively the flow of auditory and visual information into the midbrain. The topographic nature of the visual input into the OT is emphasized by individual arrows which indicate some of the spatial locations from which visual information arrives at the OT. Here, R10° (L10°) represents the location in the visual field that is 10° to the right (left) of the central azimuthal position. R: rostral, c: caudal, m: medial, l: lateral. Adapted from (DeBello *et al.*, 2001).

associations by displacing the visual input along the azimuth by a predetermined amount (Brainard and Knudsen, 1998). For instance, when an owl fitted with spectacles that produce a 10° rightward shift looks straight ahead, it receives visual information centred at 10° to the left, rather than at 0°, or straight ahead. As a result, whereas a normal owl can visually localize a sound-source placed at 0° by looking straight ahead, one fitted with a right-shifting prism has to learn to rotate its head to the right by the appropriate amount, in order to localize the same source.

It has been found that owls less than 200 days old display a remarkable ability to behaviourally adapt to prism experience—up to a 23° shift—in about seven weeks. Removal of prisms after adaptation results in gradual recovery of normal orienting behaviour. On the other hand, adult owls adapt poorly to prism experience (Knudsen, 1998). Interestingly, shifted-back juvenile owls that are subsequently fitted with prisms as adults do show adaptability, but only up

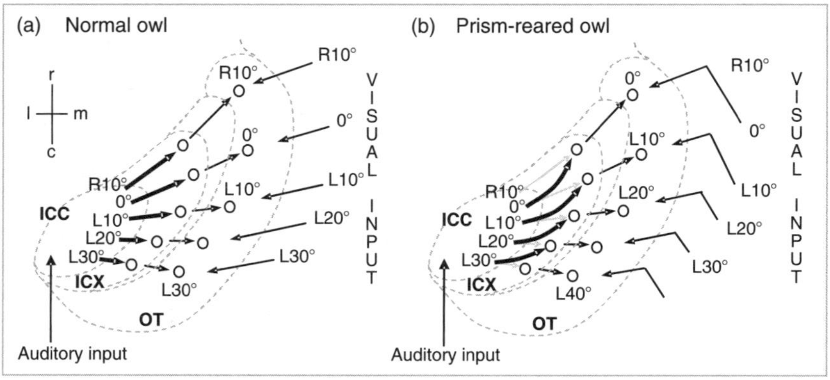

Fig. 4.2 Auditory localization plasticity in prism-reared owls. Adaptation to prism experience in the auditory localization behaviour of juvenile owls has been found to occur through activity-dependent architectural modifications of the topographic representation of auditory space. It has been found that axonal growth and synaptogenesis occur only between the ICC and ICX layers in the midbrain (DeBello *et al.*, 2001; Feldman and Knudsen, 1997). As a result, while the auditory tuning of neurons in the ICC remains unchanged, that of the ICX (and consequently the OT) neurons shifts by the amount of visual displacement induced by the prisms (Brainard and Knudsen, 1993, 1995). (a) Schematic representation of a section through the right midbrain of a normal barn owl showing the default anatomical projections (arrows) between different layers. Auditory (ITD) information arriving at the ICC and visual information arriving at the OT are shown. The open circles represent individual neurons. The text above each circle represents the azimuthal centre of the auditory receptor field of that neuron. The text adjacent to the arrows represents the centre of the spatial location from which auditory or visual information arrives to the midbrain. The thick arrows between ICC and ICX represent the integration of tonotopic information across various frequency channels of the ICC. (b) Schematic representation of a section through the right midbrain of a prism-reared barn owl. The auditory input remains unchanged, while the visual input is shifted as a result of prism fitting (depicted with arrows that are bent). New anatomical projections between the ICC and ICX that develop in response to prism-rearing are shown with thick curved arrows. Text above the neurons in the OT highlights the shift in the auditory tuning of these neurons. For instance, the tuning of the topmost neuron in the OT shifts from R10° to 0°. Here, R10° (L10°, represents the location in the visual field that is 10° to the right (left) of the central azimuthal position. R: rostral, c: caudal, m: medial, l: lateral. Adapted from DeBello *et al.* (2001).

to the adjustments they had originally displayed (Knudsen, 1998). Recent evidence (Linkenhoker and Knudsen, 2002) also shows that the capacity for behavioural plasticity in adult owls can be increased with incremental training, that is, where the adults experience prismatic shifts in small increments.

Detailed investigations into the mechanisms mediating this behavioural adaptation have yielded several insights. Particularly interesting is the fact that learning is mediated by structural changes—the growth of new axons and the formation of new synapses—in the neural architecture of prism-fitted birds (see Figure 4.2).

We are developing a model of this system to understand the computational mechanisms through which different aspects of the observed architectural adaptation can be implemented in a biologically feasible manner. Box 4.1 discusses the model, and preliminary results show that a Hebb-like learning rule can account for many of the experimental details. The model makes predictions about the site of an important (inhibitory) component of plasticity, and about the biological signals that can trigger structural change.

Box 4.1 Modelling plasticity in the auditory localization system (ALS) of barn owls

We implemented a simple connectionist model of the auditory localization network in barn owls using firing-rate coding neurons. Figure 4.3 shows the architecture of the model based on published details of the anatomical connections in the midbrain localization pathway (Hyde and Knudsen 2000; Knudsen 2002). The different layers are the ICC, ICX, OTD (deep layers of the OT), OTM (medial layers of the OT), and OTS (superficial layers of the OT). Neurons in the ICX are either excitatory (ICXE) or inhibitory (ICXI). Individual neurons are shown as circles, and their point-to-point projections are indicated by arrows. We assume without loss of generality that every degree in the visual (and auditory) field is coded for by a separate neuron in any given layer. The output of each neuron is a nonlinear function of the weighted sum of its inputs. The driving assumption of our modelling effort is that all the default structures and pathways already exist and that the phenomenon of interest is the plasticity between the ICC and ICX layers in response to prismatic shifts in the visual field. The *ab initio* development of ITD maps has been modeled in literature (for instance, see Kempster *et al.*, 2001), but we do not focus on this. For simplicity, we omit the frequency dependence of ICC neuronal tuning in this model, and assume without loss of generality that the ITD tuning of neurons is all-or-none corresponding to the best ITD.

Past modelling efforts have either used a foveation-based error signal to drive learning (Gelfland *et al.*, 1988; Pouget *et al.*, 1995; Rucci *et al.*, 1997) which has since been shown to be incorrect, or have primarily relied

Box 4.1 *(continued)*

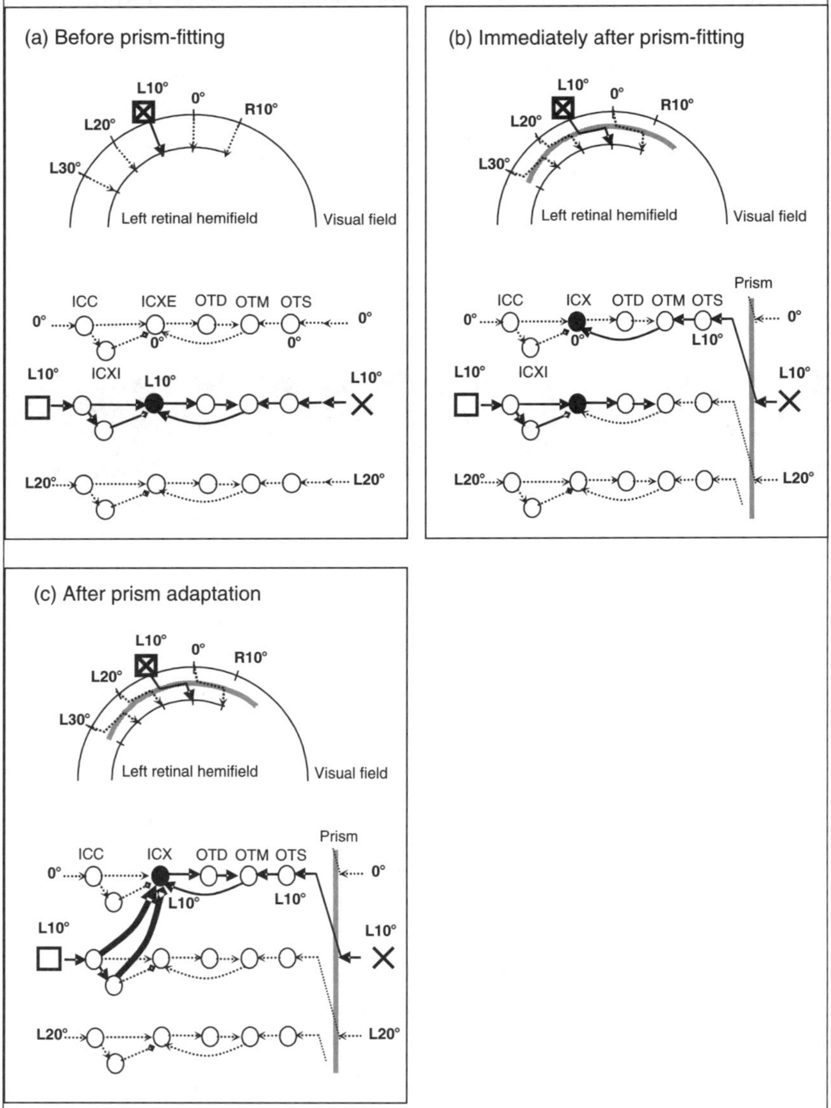

Fig. 4.3 Model of the midbrain auditory localization system in barn owls. Shown are the three stages in the adaptation of the system to prism experience. In the top half of each panel, the outer semi-circle is a cartoon representation of the audiovisual field experienced by an owl. The inner arc depicts the left retinal hemifield. Representative locations in the audiovisual field are represented in degrees. An audio stimulus is represented as an open square, a visual stimulus as an 'X', and a combined audiovisual stimulus as a square and an 'X'. In all cases, an audio-visual stimulus is presented to the owl at L10° (i.e., the location in space that is at 10° to the left of the central azimuthal position).

Box 4.1 *(continued)*

Fig. 4.3 Caption Continued The bottom half of each panel shows the topographic nature of the information flow (dotted and continuous arrows) through different layers in a model of the midbrain. Each column of circles represents a different layer of neurons in the midbrain, and each row represents the point-to-point pathway that processes information from one spatial location. The inhibitory neurons in the ICX project to the excitatory neurons in the ICX and this inhibitory connection is shown by an arrow with a diamond-shaped head between an ICXI and an ICXE neuron. Auditory input enters the model at the left through ICC neurons, and visual input at the right, through OT neurons. (a) Here, a colocalized audiovisual input is presented to a normal owl at L10° (shown in the top half). The default anatomical connections and flow of information are shown in the bottom half. We see that audio input arrives at the neuron encoding for L10° in the ICX (filled circle), and the visual input arrives at the same ICX neuron via a back projection from the corresponding OT neuron. The pathway activated by this input is represented with continuous arrows, while non-active pathways are shown with dotted arrows. As mentioned in the main text, such a stimulus produces a firing response in the ICX neuron about 10ms after its presentation. (b) In this panel, the same audio-visual input as in (a) is presented to a juvenile owl immediately after prism-fitting. The prism is schematically shown in the top half of the panel as a thick arc, which refracts all incoming visual input, and shifts it to the right by 10°. The bottom half shows auditory input arriving as in a normal owl, and being processed by the ICX neuron in the middle row (filled circle). However, due to prism-fitting visual information from L10° is now processed by the neuron in the OTS that had previously processed visual information from 0° (compare with panel a). The prism is shown here as a thick line. The point-to-point anatomical projections imply that the downstream neurons that subsequently process this visual information are also different from those in a normal owl. We see that the visual signal eventually gets back-projected to the ICX neuron in the top row (shown as a filled circle), whose auditory tuning corresponds to 0°. Thus, there is a mismatch between the processing of auditory and visual information at the ICX. It has been found that in this case, the ICX neuron in the top row, which receives only the visual component of the bimodal stimulus, fires approximately 90 ms after stimulus presentation (Gutfreund *et al.*, 2002). This delayed firing of the ICX neuron constitutes the error signal that subsequently drives plasticity in this system. (c) In this panel, the same audio-visual stimulus as before is presented to the prism-fitted juvenile owl after it has displayed behavioral adaptation to prism experience. New anatomical projections that were observed between the ICC and ICX (DeBello *et al.*, 2001; Feldman and Knudsen, 1997) are shown in the bottom half of this panel by thick arrows. These new projections have shifted the auditory tuning of the ICX neurons (but not that of the ICC neurons), and as a result, the mismatch in the audiovisual information processing in the ICX has been eliminated. The ICX neuron that now processes the audiovisual input is in the top row (shown by a filled circle), and its new auditory tuning corresponds to L10°.

Box 4.1 *(continued)*

on reward-based learning schemes (Pouget *et al.*, 1995; Rucci *et al.*, 1997) about which little is known in the owl's auditory localization system. Although Rosen *et al.* (1994) did not use a reinforcement learning scheme, they implemented learning via the back-propagation algorithm. The learning rule we use is inspired by the classical Hebbian scheme. We move away from reward-based methods and a global error signal, and implement learning using a template-based error signal. This is based on recent evidence showing that the teaching or error signal that drives plasticity in this system is a visually based topographic template signal (Hyde and Knudsen, 2001). It has also been recently established that visual input from the OT does indeed project to the ICX and that this back-projecting error signal which is responsible for the calibration of the map is encoded in the selective and delayed firing of ICXE neurons in response to the presentation of a co-localized audiovisual stimulus to a prism-shifted system (Gutfreund *et al.*, 2002). In such a set-up, ICXE neurons receive either the visual or the auditory component of the input (see Figure 4.3). Under normal circumstances, a co-localized bimodal stimulus produces a response in the corresponding ICX neuron about 10 ms after the stimulus is presented, but significantly before (about 80 ms before) the response in the prism-shifted case (Gutfreund *et al.*, 2002). In our model, we detect the error signal by checking for ICX activity 90 ms after the presentation of an audiovisual stimulus. Weights are reduced between ICC neurons that do not fire and ICXE (and ICXI) neurons that do fire as a result of the back-projecting error signal from the OT. If the error occurs repeatedly for a sufficiently long period of time (implying a systematic error in encoding information), new synapses are formed, that is, the population of the ICC-ICXI and ICC-ICXE weight matrices is increased. Once this happens, the learning rule increases weights between ICC neurons that fire and ICXE (and ICXI) neurons that also fire soon after (indicating the arrival of auditory input). The time course of weight evolution is implemented to be sigmoidal, and this prevents an indefinite increase in weights caused by the classical Hebbian rule. The results of this scheme match those from experiments in that the ITD tuning curves of the ICX neurons shift in the adaptive direction after training, and we are able to capture the essence of the learning exhibited by the barn owls. Current work is looking into relaxing some of the assumptions made at the beginning, and at incorporating more details into the model.

Structural plasticity and representation construction

The initial topographic organization of the auditory localization system suggests that structural plasticity may be necessary for the owl to adapt to experimentally induced visual displacement. Indeed, a temporally restricted developmental window for structural plasticity may be the reason adult owls are unable to adapt beyond a few degrees of shift (Brainard and Knudsen, 1998). An interesting question in this context is, what are the implications of this capacity for architectural change to the representational properties of the network in the context of the highly constrained neurobiological framework?

Investigating the representational consequences of architectural plasticity necessitates the use of appropriate quantifiers of representational complexity. One way to quantify the 'ability' or 'complexity' of a network is in terms of its memory capacity (Poirazi and Mel, 2001), defined using a function-counting approach. We apply a similar approach to the midbrain auditory localization network in the barn owl by treating the auditory and visual inputs as the *input* to the network, and the activities of the ICX neurons as its *output*. If we assume for simplicity that every neuron codes uniquely for one degree in the visual (and auditory) field, and that each neuron in one layer projects uniquely to the corresponding neuron in another, then a simple calculation yields that the initial network architecture implements one input–output function, whereas the post-learning architecture (that results after a one degree prism shift) implements a qualitatively different function. That is, structural plasticity has permitted an increase in the repertoire of localization functions that the network can compute from one to two, while satisfying topographic projection constraints. According to this calculation scheme, the increase in memory capacity (or function complexity) possible in response to an n-degree prismatic shift in juvenile owls is n. We note here that the ability to grow introduces extra degrees of freedom in the input–output transformations in a manner that is consistent with the shift in the resulting output. That is, if the output units are considered to be linear for simplicity, then the input–output function between a pair of layers is a linear transformation (or a matrix), and the point-to-point nature of the anatomical projections constrain all off-diagonal elements of this matrix to be zero. Synaptic weight modifications just scale the diagonal elements, whereas growth permits the population of appropriate off-diagonal elements resulting in qualitatively different functions. No amount of diagonal scaling can reproduce the latter effect. As a result, if there is no growth but only synaptic efficacy change, the memory capacity of the network and hence its representational complexity remain constant. This observation is consistent with the hypothesis that a lack of the ability to grow in adult circuits may be the reason adult owls (>200 days old) are unable to adapt to large visual shifts.

The general lesson from this model system is that environmentally derived instructive signals help regulate architectural plasticity and that this form of plasticity underlies abrupt qualitative change in the representational ability of a network. Further, this representational change typically increases the functional complexity of a network. Viewed from the perspective of learning theory, the initial network represents the hypothesis space that is explored while learning from examples and architectural plasticity allows the developing owl to go beyond this original hypothesis space to learn novel functions that are originally outside of this space. In contrast, the mature owl appears not to possess the same degree of architectural plasticity, and so is limited in terms of learning capacity. This highlights the fundamental role of architectural plasticity as a form of constructive learning and illustrates why it is fundamentally different from synaptic weight change, which only implements a search within a given hypothesis space. Hence architectural plasticity is a powerful and indispensable ally in representation construction and it is likely a general theme of neurobiological development in many species, including humans, a species to which we now turn.

Evidence from modelling cognitive development in children

One way to assess the importance of constructive growth in cognitive development is to compare models whose networks are allowed to grow to models containing static networks. This comparison is facilitated by the fact that the principal difference between two of the most popular neural algorithms for modelling cognitive development is that one of them incorporates growth and the other does not (Shultz, 2003). Both back-propagation (BP) and cascade-correlation (CC) involve multilayered feedforward networks that learn by adjusting their connection weights to reduce error, defined as the discrepancy between actual and target outputs. BP networks are typically designed by a programmer and retain their initial structure throughout learning. In contrast, CC networks begin with a minimal topology containing only inputs and outputs defined by a programmer, and then proceed to add hidden units as needed to master the training problem.

Box 4.2 describes the CC algorithm, which has been successfully applied to a wide variety of phenomena in cognitive development in children (Shultz, 2003). There are presently three domains of cognitive development to which both BP and CC networks have been applied. Here we briefly review the simulations in these domains in order to examine the effects of growth on simulation success. The three relevant domains are the balance scale, the integration of velocity, time, and distance cues for moving objects, and age changes in sensitivity to correlations vs. features in infants' category learning. Note that in none of these domains do we know about the actual brain circuits involved. Nor are the BP and CC algorithms considered to be realistic implementations of the details of real brain circuits. They are instead highly abstracted, functional algorithms, albeit ones that are inspired by some general neurophysiological constraints (Shultz, 2003).

Box 4.2 The cascade-correlation (CC) algorithm

There are two phases in CC learning—the input phase and the output phase. Training typically begins in output phase by using the so-called delta rule to adjust the weights entering output units in order to reduce network error. When error reduction stagnates for a set number of epochs or when the problem has not been mastered within another, larger certain number of epochs, CC switches to input phase. An epoch is a pass through the training patterns.

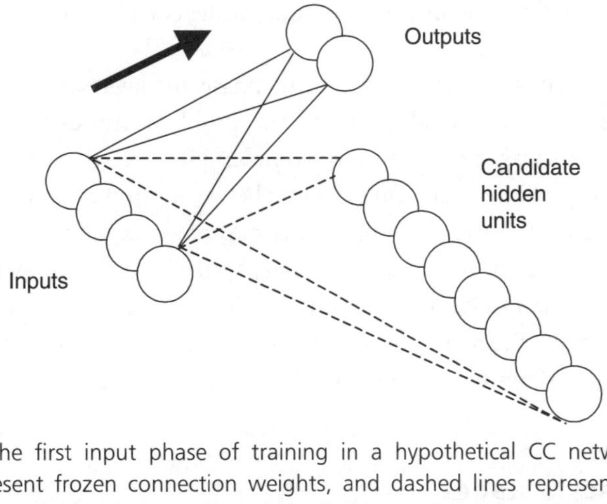

Fig. 4.4 The first input phase of training in a hypothetical CC network. Solid lines represent frozen connection weights, and dashed lines represent trainable connection weights (Shultz, 2003, adapted with permission from MIT Press).

In the input phase, the task is no longer to reduce error, but rather to reconceptualize the problem by recruiting a useful hidden unit downstream of the existing units but upstream of the output units. A pool of typically eight candidate hidden units, usually equipped with sigmoid

Box 4.2 *(continued)*

activation functions, is evaluated for recruitment (see Fig. 4.4). This evaluation consists of training randomly initialized weights from the inputs and any existing hidden units to these candidates, with the goal of maximizing a modified correlation between candidate activation and network error. In other words, the CC algorithm in input phase is looking for a new hidden unit whose activation patterns track the network's current error, becoming either very active or very inactive with error fluctuations—in short, a candidate unit which is sensitive to the major difficulties that the network is currently having. The same delta rule is used here in input phase, but instead of minimizing error as in output phase, it is used to maximize correlations.

When these correlations stagnate over a certain number of input-phase epochs, the candidate with the highest absolute correlation with network error is selected and installed into the network, and the less successful candidates are discarded. This implements a kind of proliferation and pruning of units and connections that is characteristic of selection models of brain development (Changeux and Danchin, 1976). The new recruit is initialized with randomized output weights of positive values if the correlation with error was negative or with negative values if the correlation with error was positive.

At this point, CC returns to output phase to determine how to best utilize the new conceptualization of the problem afforded by its latest recruit. In this way, new and more complex interpretations build on top of older, simpler interpretations. Cycling between input and output phases continues until the training problem is mastered. CC can be interpreted as implementing the processes of synaptogenesis and neurogenesis that our review shows to be at least partly under the control of learning. CC also implements the hierarchical, cascaded progression now thought to be characteristic of brain development (Thivierge, Rivest and Shultz, 2003).

Balance-scale stages

The balance-scale problem involves presenting a child with a rigid beam balanced on a fulcrum (Siegler, 1976, 1981). The beam is equipped with several pegs spaced at regular intervals to the left and right of the fulcrum. The experimenter typically places some number of identical weights on one peg on the left side and some number of weights on one peg on the right side. While supporting blocks prevent the beam from moving, the child is asked to predict which side of the beam will descend, or whether the scale will remain balanced, once the supporting blocks are removed.

The 'rules' that children use to make balance-scale predictions can be diagnosed by presentation of six different types of problems. Three problems are relatively simple in the sense that one cue (either weight or distance from the fulcrum) predicts the outcome because the other cue does not differ from one side of the scale to the other. Three other problems are relatively complex in that the two relevant cues conflict with each other, with weight predicting one outcome and distance predicting the opposite outcome. For simple or complex types, there are three possible outcomes in which the direction of the tip is predicted by weight information or by distance information or the scale remains balanced. The pattern of a child's predictions across these six different problem types can be used to diagnose the 'rule' the child is apparently using to make predictions. Of course, in a connectionist network a rule is not implemented as a symbolic if-then structure, but is rather an emerging epiphenomenon of the network's topology and connection-weight values.

There is a consensus in the psychological literature that children progress through four different stages of balance-scale performance (Siegler, 1976, 1981). In stage 1, children use weight information only, predicting that the side with greater weight will descend or that the scale will balance when the two sides have equal weights. In stage 2, children continue to use weight information but begin to use distance information when the weights are equal on each side. In stage 3, weight and distance information are used about equally, but the child guesses when weight and distance information conflict on the more complex problems. Finally in stage 4, children respond correctly on all types of problems, whether simple or complex.

In one of the first connectionist simulations of cognitive development McClelland and Jenkins found that a static BP network with two groups of hidden units segregated for either weight or distance information progressed through the first three stages of the balance scale and even into the fourth stage (McClelland and Jenkins, 1991). However, the network never settled into stage 4 performance, instead cycling between stage 3 and stage 4 functioning for as long as the programmer's patience lasted. Further simulations indicated that BP networks could settle into stage 4, but only at the cost of missing stages 1 and 2 (Schmidt and Shultz, 1991). There seems to be no way for BP networks to capture both consistently mature stage 4 performance and earlier progressions through stages 1 and 2. In contrast, the first CC model of cognitive development naturally captured all four stages of balance-scale performance, and did so without requiring any hand-designed segregation of hidden units (Shultz *et al.*, 1994). Capturing balance-scale stages is not a trivial matter, as witnessed by the shortcomings of earlier symbolic-rule-learning models (Langley, 1987; Newell, 1990).

Both BP and CC models are able to capture naturally the other psychological regularity of balance-scale development, the torque-difference effect. This is the tendency for problems with large absolute torque differences (from one side of the scale to the other) to be easier for children to solve, regardless of their current stage (Ferretti and Butterfield, 1986). Connectionist feed-forward networks produce perceptual effects like torque difference whenever they learn to convert quantitative input differences into a binary output judgment. The bigger the input differences—in this case, from one side of the fulcrum to the other side—the clearer the hidden-unit activations, and the more decisive the output decisions. Symbolic rule-based models are unable to naturally capture the torque-difference effect because symbolic rules typically don't care about the amount of differences, only the direction of differences.

The ability of neural-network models to capture stages 1 and 2 on the balance scale is due to a bias towards equal-distance problems in the training patterns. These are problems in which the weights are placed equally distant from the fulcrum on each side. This bias forces the network to first emphasize weight information at the expense of distance information because weight information is much more relevant to reducing prediction errors. Once weight information is successfully dealt with, then the network can turn its attention to distance information, particularly on those problems where distance varies from one side of the fulcrum to the other. In effect, the network must find the particular region of connection-weight space that allows it to emphasize the numbers of weights on the scale and then move to another region of weight space that focuses on the multiplication of equally important weight and distance information. It requires a powerful learning algorithm to make this move in connection-weight space. Apparently a static BP network, once committed to using one source of information in stage 1, cannot easily find its way in weight space to the stage 4 region merely by continuing to reduce error. A constructive algorithm, such as CC, has an easier time with this move because each newly recruited hidden unit effectively changes the shape of connection-weight space by adding a new dimension. This new dimension affords a path on which to move towards the stage-4 region. More colloquially, BP cannot have its cake and eat it too, but CC, due to its dynamic increases in computational power, can both have and eat its cake, essentially by continuing to bake.

Integration of velocity, time, and distance cues

In Newtonian physics, the velocity of a moving object is the ratio of distance traveled to the time of the journey: velocity = distance/time. Algebraically then, distance = velocity x time, and time = distance/velocity. Some of the cleanest evidence on children's acquisition of these relations was collected by Wilkening (1981), who asked children to predict one dimension (e.g., time) from knowledge of the other two (e.g., velocity and distance). For example, three levels of velocity information were represented by the locomotion of a turtle, a guinea pig and a cat. These three animals were described as fleeing from a barking dog, and a child participant was asked to imagine these animals running while the dog barked. The child's task was to infer how far an animal would run given the length of time the dog barked, an example of inferring distance from velocity and time cues. CC networks learning similar tasks typically progress through an equivalence stage (e.g., velocity = distance), followed by an additive stage (e.g., velocity = distance − time), and finally the correct multiplicative stage (e.g., velocity = distance/time) (Buckingham and Shultz, 2000). Some of these stages had been previously found with children, and others were subsequently confirmed as predictions of our CC simulations. As with children in psychological experiments, our CC networks learned to predict the value of one dimension from knowledge of values on the other two dimensions.

We diagnosed rules based on correlations between network outputs and the various algebraic rules observed in children. To be diagnosed, an algebraic rule had to correlate positively with network responses, account for most of the variance in network responses, and account for more variance than any other rules did. As with rules for solving balance-scale problems, these rules are epiphenomena emerging from patterns of network connection weights. For velocity and time inferences, CC networks first acquired an equivalence rule, followed by a difference rule, followed in turn by the correct ratio rule. Results were similar for distance inferences, except that there was no equivalence rule for distance inferences. In making distance inferences, there is no reason a network should favour either velocity or time information because both velocity and time vary proportionally with distance.

A shift from linear to non-linear performance occurred with continued recruitment of hidden units. Linear rules include equivalence (e.g., time = distance), sum (e.g., distance = velocity + time), and difference (e.g., velocity = distance − time) rules, whereas non-linear rules include product (e.g., distance = velocity x time) and ratio (e.g., time = distance/velocity) rules.

Because the sum and difference rules in the second stage are linear, one might wonder why they require a hidden unit. The reason is that networks without a hidden unit are unable to simultaneously encode the relations among the three dimensions for all three inference types. In distance inferences, distance varies directly with both velocity and time. But in velocity inferences, distance and time vary inversely, and in time inferences, distance and velocity also vary inversely. Networks without hidden units are unable to encode these different relations. The first-recruited hidden unit differentiates distance information from velocity and time information, essentially by learning connection weights with one sign (positive or negative) from the former input and opposite signs from the latter inputs. This weight pattern enables the network to consolidate the different directions of relations across the different inference types.

In contrast to constructive CC networks, static BP networks seem unable to capture these stage sequences (Buckingham and Shultz, 1996). If a static BP network has too few hidden units, it fails to reach the correct multiplicative rules. If a static BP network has too many hidden units, it fails to capture the intermediate additive stages on velocity and time inferences. Our extensive exploration of a variety of network topologies and variation in critical learning parameters suggests that there is no static BP network topology that can capture all three types of stages in this domain. Even the use of cross-connections that bypass hidden layers, a standard feature of CC, failed to improve the stage performance of BP networks. There is no apparent way to get BP to cover all three stages because the difference between underpowered and overpowered BP networks is a single hidden unit. Thus, we conclude that the ability to grow in computational power is essential in simulating stages in the integration of velocity, time, and distance cues.

Age changes in infant category learning

Our final simulation comparison concerns age changes in sensitivity to correlations vs. features in category learning by infants. Using a stimulus-familiarization-and-recovery procedure to study categorization, Younger and Cohen found that 4-month-olds process information about independent features of visual stimuli, whereas 10-month-olds are able to abstract relations among those features (Younger and Cohen, 1983, 1986). Such findings relate to a long-standing controversy about the extent to which perceptual development involves integration (Hebb, 1949) or differentiation (Gibson, 1969) of stimulus information. A developing ability to understand relations among features suggests that perceptual development involves information integration, a view compatible with constructivism.

Infants are assumed to construct representational categories for repeated stimuli, ignoring novel stimuli that are consistent with a category, while

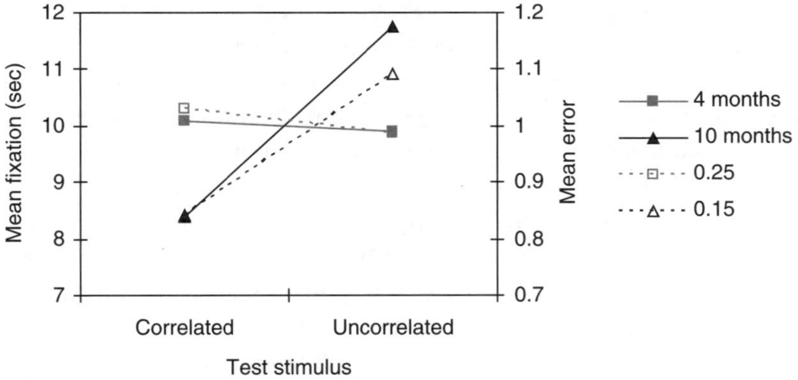

Fig. 4.5 Mean fixation of key test stimuli in infants (solid lines) from Younger and Cohen (Younger and Cohen, 1986, reprinted with permission of the Society for Research in Child Development), and mean CC network error at two levels of score-threshold (dashed lines) from Shultz and Cohen (Shultz and Cohen, 2004, adapted with permission).

concentrating on novel stimuli that are not members of a category (Cohen and Arthur, 1983; Younger and Cohen, 1985). After repeated presentation of visual stimuli with correlated features, 4-month-olds recovered attention to stimuli with novel features more than to stimuli with either correlated or uncorrelated familiar features (Younger and Cohen, 1983, 1986). In contrast, 10-month-olds recovered attention to both stimuli with novel features and familiar uncorrelated features more than to stimuli with familiar correlated features. Mean fixation time in seconds is shown for the key interaction between age and correlated vs. uncorrelated test stimuli, plotted in solid lines in Figure 4.5 against the left-hand *y*-axis. This pattern of recovery of attention indicates that young infants learned about the individual stimulus features, but not about the relationships among features, whereas older infants in addition learned about how these features correlate with one another.

These infant experiments were recently simulated with CC encoder networks (Shultz and Cohen, 2004). Encoder networks have the same number of input units as output units, and their job is to reproduce their input values on their output units. They learn to do this reproduction by encoding an input representation onto their hidden units and then decoding that representation onto output units. In this fashion, encoder networks develop a recognition memory for the stimuli they are exposed to. Network error can be used as an index of stimulus novelty. Both BP and CC encoders were applied to the infant data (Shultz and Cohen, 2004). In CC networks, age was implemented by varying the score-threshold parameter: 0.25 for 4-month-olds and 0.15 for

10-month-olds. This parameter governs how much learning occurs because learning stops only when all network outputs are within score-threshold of their target values for all training patterns. It was assumed that older infants would learn more from the same exposure time than would younger infants. Mean network error for the key interaction between age and correlated vs. uncorrelated test stimuli is plotted with dashed lines in Figure 4.5 against the right-hand *y*-axis. On the assumption that network error reflects stimulus novelty (and thus infant interest), the CC networks closely matched the infant data—more error to the uncorrelated test stimulus than to the correlated test stimulus only at the smaller score-threshold of 0.15.

In sharp contrast, a wide range of static BP networks could not capture this key interaction. Our BP simulator was modified to use a score-threshold parameter to decide on learning success, just as in CC. The same coding scheme used in the CC simulations was also used here. A wide variety of score-threshold values were explored in a systematic attempt to cover the infant data. BP networks were equipped with three hidden units, the typical number recruited by CC networks. BP network topology was varied to explore the possible roles of network depth and the presence of cross connections in simulation success. Both flat and deep BP networks were tried, both with and without the cross-connection weights native to CC. There were six input and six output units in BP networks just as there were in CC networks. In the BP networks, there was never lower error on correlated than on uncorrelated test items, the signature of effective correlation detection. It is difficult to prove that any algorithm cannot in principle cover a set of phenomena, but we certainly gave BP a fair shake, running 9 networks in each of 80 simulations (2 familiarization sets × 10 score-threshold levels × 4 network topologies).

Unlike previous psychological explanations that postulated unspecified qualitative shifts in processing with age, our computational explanation based on CC networks focused on quantitatively deeper learning with increasing age, a principle with considerable empirical support over a wide range of ages and experiments (e.g., Case *et al.*, 1982). CC networks also generated a crossover prediction, with deep learning showing a correlation effect, and even more superficial learning showing the opposite—a similarity effect, in the form of more error to (or interest in) the correlated test item than to the uncorrelated test item. This is called a *similarity* effect because the uncorrelated test item was most similar to those in the familiarization set. This simulation predicted that with a single, suitable age group, say 10-month-olds, there would be a correlation effect under optimal learning conditions and a similarity effect with less than optimal familiarization learning. Tests of this prediction found that 10-month-olds who habituated to the training stimuli looked longer at the uncorrelated than the correlated test stimulus, but those who did not habituate did just the opposite, looking longer at the correlated than the uncorrelated test stimulus (Cohen and Arthur, unpublished).

In addition to covering the essential age by correlation interaction, CC networks covered two other important features of the infant data: discrimination of the habituated from the correlated test stimulus via the presence of two additional non-correlated features, and more recovery to test stimuli having novel features than to test stimuli having familiar features.

Although the BP simulations did not capture the infant data, they were useful as control conditions in establishing why CC networks were successful. Namely, the growth of cascaded networks with cross-connection weights seems critical to capturing detection of correlations between features and the developmental shift to this ability from earlier feature-value learning.

Because CC is sufficiently abstracted from the low-level details of actual neural circuits, it is not clear whether its recruitment of hidden units corresponds to synapto- or neurogenesis. This issue concerns whether candidate units already exist somewhere else in the system or whether they are created fresh before recruitment. From a sufficiently abstract computational view, it seems not to matter. But if CC does implement synaptogenesis, then it assumes establishment of fresh synapses rather than adding duplicate synaptic connections between units.

Conclusions

The chief biological constraint implemented in our models is that of letting networks grow while they learn. We allowed our networks to grow by forming new connections or by recruiting new hidden units. The modeling revealed a number of computational advantages for network growth, all of which made for more realistic coverage of both animal and human development.

The work reviewed here indicates that brain networks in a variety of species clearly grow as they learn. There is good evidence of synaptogenesis in birds and mammals in the cerebellum, hippocampal formation, and sensory and motor cortices. There is consensus that neurogenesis occurs in the olfactory bulb and dentate gyrus of a variety of mammals including monkeys, but published evidence of cortical neurogenesis is still somewhat controversial. Evidence indicates that both synapto- and neurogenesis increase with pressures and opportunities to learn, and that they play an important functional role in learning.

Computational analysis shows that constructive learning algorithms offer several advantages over static learning algorithms. Constructive algorithms can learn any learnable problem in realistic time by reducing bias while keeping variance low, thus avoiding the bias-variance trade-off that plagues static learners. Unlike static learners, constructive algorithms can also escape from Fodor's paradox about not being able to learn anything genuinely new. New representations constructed by growing networks can be qualitatively different than anything they possessed before. Furthermore neural-network algorithms that grow offer a precise and convincing way to implement constructivist accounts of cognitive development wherein new knowledge builds on earlier knowledge. Finally computational arguments show it is unlikely that evolution has prepared brain networks in human children for all of the learning problems to which they might eventually be exposed. It is more likely that brain networks are constructed as novel problems arise.

Recent investigations of sound localization in the barn owl provide a clear and compelling example of learning-directed brain growth, complete with evidence of synaptogenesis in the auditory topographic map in the inferior colliculus when juvenile barn owls are fitted with optical prisms that shift the visual field across the azimuth. This qualitative change in neural circuitry and sound localization was modeled with a Hebbian-like learning rule that was allowed to adjust old synapses and form new ones in response to the mismatch between auditory and visual information.

Implementing network growth in simulations of three different domains of psychological development in children produced superior data coverage when contrasted with comparable static networks. Network growth offers a number of computational advantages in covering these phenomena including the building of more complex knowledge on top of simpler knowledge, superior ability to escape from local minima during error reduction, and deeper learning of training problems.

The simulation of infant category learning predicted a correlation effect under optimal learning conditions and a similarity effect with less than optimal learning. There is already some confirmation of this prediction in that infants who habituated to training stimuli showed a correlation effect, but those who did not habituate showed the opposite, similarity effect.

An interesting general hypothesis raised by our computer simulations is that the function of learning-driven neural growth is to increase computational power so that learning tasks can be mastered. Precisely how this hypothesis might be tested in future brain research remains to be worked out.

It is noteworthy that these computational constraints can operate in a variety of learning paradigms, including both Hebbian learning and multilayered feedforward CC networks. As used here, some of the differences between these learning paradigms became somewhat blurred. Typically considered as an unsupervised learner, the Hebb rule was here modified to use the discrepancy between visual and auditory information to produce an error signal. Typically considered as a supervised learner, CC can often be construed as not requiring a human teacher to supply target outputs. For example, CC networks learn about balance-scale problems and about velocity, time and distance relations by computing the discrepancy between what they predict will happen and what actually happens (observations of the natural environment). Likewise, CC encoder networks (such as those used in our simulations of category learning in infants) use the stimulus itself as the target output, compute error as the discrepancy between input and output activations, and in effect implement an unsupervised learning scheme.

Although we emphasized network growth in this chapter, it is also clear from neuroscience research that brains discard little-used neurons and synapses. In a drive towards even greater biological realism, it would be worth experimenting with networks that both grow and prune. Some preliminary work along these lines has found that pruning of CC networks during growth and learning improves learning speed, generalization, and network interpretability (Thivierge *et al.*, 2003). This would appear to be yet another case of a biological constraint offering computational advantages.

The bottom line is that there is a coherent alternative to the instantaneity hypothesis that we described in the Introduction. This constructivist alternative takes both development and learning seriously, and embraces the non-instantaneous features of development as fundamental to the understanding of development. As we have illustrated in this chapter, this developmental trajectory of passing through increasingly complex representational and processing capacities is both biologically realistic and computationally attractive in terms of the acquisition capacities it confers on a developing system. Models of psychological development that fail to grow come up short on all these counts. Letting networks grow and providing a role for environmentally-derived instructive signals to participate in this growth is a major departure in the study of cognitive development.

Acknowledgements

The research discussed here was supported by NSF grant 0093757 (Steven R. Quartz and Shreesh P. Mysore), and a grant from the Natural Sciences and Engineering Research Council of Canada (Thomas R. Shultz). We are grateful to Evan Balaban, Frederic Dandurand, Karim Nader, Yoshio Takane and J. P. Thivierge for helpful comments on earlier drafts.

Glossary

Basal CA1 dendrites The basal dendrites of neurons in a region called CA1 that is part of the hippocampus.

Bias-variance trade-off In formal learning theory, a quantification of the generalization ability of a learner, or equivalently, of its generalization error, is achieved statistically by defining it to be the sum of two components—bias squared and variance. Bias is defined as the distance of the best hypothesis in the space from the target concept, whereas variance is the distance of the current hypothesis from the target. Restricting the hypothesis space tends to reduce variance while increasing bias (the target concept may be far away from the few hypotheses in the space). On the other hand, expanding the hypothesis space increases variance but also increases the likelihood that the target hypothesis is included in the space (low bias). The trade-off that occurs when trying to minimize the generalization error is called the bias-variance trade-off.

Delay or classical eye-blink conditioning This is similar to trace conditioning in that the animal is trained to learn the association between an unconditioned stimulus and an aversive stimulus. The main difference is that the aversive stimulus comes on while the unconditioned stimulus is still on.

Neocortex Evolutionarily, this is the most recent part of the brain, and it is associated with higher cognitive functions in humans. Hence evidence of neurogenesis in this area in macaques has come as an unexpected finding, with potentially far-reaching implications to cognitive development, and to stem-cell research.

Polynomial time The complexity of a computational problem is one of the indices used to characterize and classify it. The time taken to solve the problem as a function of its size is a commonly used measure to quantify problem complexity. If the time taken to compute or solve a problem scales polynomially with its size, then it is said to be a polynomial time problem, and is considered to be tractable.

Score threshold A parameter in CC learning that governs depth of learning; learning proceeds until all output-unit activations are within score threshold of their target values.

Spatial training Typically, this involves having rats perform challenging spatial tasks on a day-to-day basis in their cages. Here, this was implemented by adding several movable 'floors' connected by ladders to the rats' cages, and by having them look for food placed at different locations on these floors. Doing this on a regular basis positively affects their performance in subsequent spatial tasks. The hippocampus is known to play in important role in spatial navigation in rats.

Torque The product of weight and distance on one side of a balance scale.

Trace eye-blink conditioning The animal is trained to learn the role of an unconditioned stimulus, like an auditory tone, in predicting an aversive stimulus that follows it. In eye-blink conditioning, the aversive stimulus is a puff of air in the eyes which causes the animal to blink. The word 'trace' refers to the fact that the aversive stimulus onsets with a delay after the unconditioned stimulus ends. The idea here is that encoding this associative relationship requires a memory 'trace' in the hippocampus.

Tuning curve (auditory or visual) of a neuron Characterization of the strength of the response of a neuron as the stimulus varies. For instance, the auditory tuning curve of an ICX neuron is a curve that represents the strength of its response as a function of the location of a sound stimulus in space. Usually, the response of a neuron peaks when the sound source is at a particular spatial location and this peak determines the 'best' stimulus location for the neuron.

Whisker barrels Clusters of neurons in layer IV in the somatosensory cortex of rodents that process information from specific whiskers. These barrels are organized topographically, in the sense that neighbouring whiskers transmit information to neighbouring barrels. Also, each whisker is associated with a unique barrel.

References

Altman J (1962). Are new neurons formed in the brains of adult mammals? *Science*, 135, 1127–1128.

Altman J and Das GD (1965). Autoradiographic and histological evidence of postnatal hippocampal neurogenesis in rats. *Journal of Comparative Neurology*, 124, 319–335.

Baum EB (1988). Complete representations for learning from examples. In Y Abu-Mostafa, ed. *Complexity in information theory*, pp. 77–98. Springer-Verlag.

Black JE, Isaacs KR, Anderson BJ, Alcantara AA, and Greenough WT (1990). Learning causes synaptogenesis, whereas motor activity causes angiogenesis, in cerebellar cortex of adult rats. *Proceedings of the National Academy of Sciences*, 87(14), 5568–5572.

Brainard MS and Knudsen EI (1993). Experience-dependent plasticity in the inferior colliculus: a site for visual calibration of the reward representation of auditory space in the barn owl. *Journal of Neuroscience*, 13, 4589–4608.

Brainard MS and Knudsen EI (1995). Dynamics of visually guided auditory plasticity in the optic tectum of the barn owl. *Journal of Neurophysiology*, 73, 595–614.

Brainard MS and Knudsen EI (1998). Sensitive periods for visual calibration of the auditory space map in the barn owl optic tectum. *Journal of Neuroscience*, 18, 3929–3942.

Buckingham D and Shultz TR (1996). Computational power and realistic cognitive development. *Proceedings of the Eighteenth Annual Conference of the Cognitive Science Society*, 507–511, Mahwah, NJ: Erlbaum.

Buckingham D and Shultz TR (2000). The developmental course of distance, time, and velocity concepts: A generative connectionist model. *Journal of Cognition and Development*, 1, 305–345.

Case R, Kurland DM, and Goldberg J (1982). Operational efficiency and the growth of short-term memory span. *Journal of Experimental Child Psychology*, 33, 386–404.

Chang PL, Isaacs KR, and Greenough WT (1991). Synapse formation occurs in association with the induction of long-term potentiation in two-year-old rat hippocampus *in vitro*. *Neurobiology of Aging*, 12, 517–522.

Changeux JP and Danchin A (1976). Selective stabilization of developing synapses as a mechanism for the specification of neuronal networks. *Nature*, 264, 705–712.

Chomsky N (1980). Rules and representations. *Behavioral and Brain Sciences*, 3, 1–61.

Cohen LB and Arthur AE (1983). Perceptual categorization in the infant. In E Scholnick (ed.) *New trends in conceptual representation*, pp. 197–220. Hillsdale, NJ: Lawrence Erlbaum Associates.

Cohen LB and Arthur AE (unpublished). The role of habituation in 10-month-olds' categorization.

Dailey ME and Smith SJ (1996). The dynamics of dendritic structure in developing hippocampal slices. *Journal of Neuroscience*, 16(9), 2983–2994.

DeBello WM, Feldman DE, and Knudsen EI (2001). Adaptive axonal remodeling in the midbrain auditory space map. *Journal of Neuroscience*, 21(9), 3161–3172.

Eriksson PS, Perfilieva E, Bjork-Eriksson T *et al.* (1998). Neurogenesis in the adult human hippocampus. *Nature Medicine*, 4(11), 1313–1317.

Feldman DE and Knudsen EI (1997). An anatomical basis for visual calibration of the auditory space map in the barn owl's optic tectum. *Journal of Neuroscience*, 17, 6820–6837.

Ferretti RP and Butterfield EC (1986). Are children's rule-assessment classifications invariant across instances of problem types? *Child Development*, 57(6), 1419–1428.

Field J, Nuir D, Pilon R, Sinclair M, and Dodwell P (1980). Infants' orientation to lateral sounds from birth to three months. *Child Development*, 51, 295–298.

Fischer M, Kaech S, Knutti D, and Matus A (1998). Rapid actin-based plasticity in dendritic spines. *Neuron*, 20(5), 847–854.

Fodor J (1980). On the impossibility of learning 'more powerful' structures. In M Piatelli-Palmarini, ed. *The debate between Jean Piaget and Noam Chomsky*, pp. 142–152. Routledge & Kegan Paul.

Gelfland JJ, Pearson JC, Spence CD, and Sullivan WE (1988). Multisensor integration in biological systems. *3rd IEEE International Symposium on Intelligent Control*, 3, 147–153, Arlington, VA.

Gibson EJ (1969). *Principles of perceptual learning and development*. New York, Appleton-Century-Crofts.

Gilbert SF (2003). *Developmental biology*. Sunderland, MA, Sinauer Associates.

Gold M (1967). Language identification in the limit. *Information and Control*, 10, 447–474.

Goldman SA and Nottebohm F (1983). Neuronal production, migration, and differentiation in a vocal control nucleus of the adult female canary brain. *Proceedings of the National Academy of Sciences*, 80(8), 2390–2394.

Gould E, Beylin A, Tanapat P, Reeves A, and Shors TJ (1999a). Learning enhances adult neurogenesis in the hippocampal formation. *Nature Neuroscience*, 2(3), 260–265.

Gould E and Gross CG (2002). Neurogenesis in adult mammals: some progress and problems. *Journal of Neuroscience*, 22(3), 619–623.

Gould E, Reeves AJ, Graziano MSA, and Gross CG (1999b). Neurogenesis in the neocortex of adult primates. *Science*, 286, 548–552.

Gross CG (2000). Neurogenesis in the adult brain: death of a dogma. *Nature Reviews Neuroscience*, 1, 67–73.

Gutfreund Y, Zheng W, and Knudsen EI (2002). Gated visual input to the central auditory system. *Science*, 297, 1556–1559.

Hebb DO (1949). *The organization of behavior*. New York, Wiley.

Hyde PS and Knudsen EI (2000). Topographic projection from the optic tectum to the auditory space map in the inferior colliculus of the barn owl. *Journal of Comparative Neurology*, 421, 146–160.

Hyde PS and Knudsen EI (2001). A topographic instructive signal guides the adjustment of the auditory space map in the optic tectum. *Journal of Neuroscience*, 21(21), 8586–8593.

Kempermann G (2002). Why new neurons? Possible functions for adult hippocampal neurogenesis. *Journal of Neuroscience*, 22(3), 635–638.

Kempermann G, Kuhn HG, and Gage FH (1997). More hippocampal neurons in adult mice living in an enriched environment. *Nature*, 386(6624), 493–495.

Kempter R, Leibols C, Wagner H, and Hemmen JLV (2001). Formation of temporal-feature maps by axonal propagation of synaptic learning. *Proceedings of the National Academy of Sciences, USA*, 98(7), 4166–4171.

Kleim JA, Freeman JH, Jr., Bruneau R *et al.* (2002). Synapse formation is associated with memory storage in the cerebellum. *Proceedings of the National Academy of Sciences*, 99(20), 13228–13231.

Knafo S, Grossman Y, Barkai E, and Benshalom G (2001). Olfactory learning is associated with increased spine density along apical dendrites of pyramidal neurons in the rat piriform cortex. *European Journal of Neuroscience*, 13(3), 633–638.

Knott GW, Quairiaux C, Genoud C, and Welker E (2002). Formation of dendritic spines with GABAergic synapses induced by whisker stimulation in adult mice. *Neuron*, 34(2), 265–273.

Knudsen EI (1998). Capacity for plasticity in the adult owl auditory system expanded by juvenile experience. *Science*, 279, 1531–1533.

Knudsen EI (2002). Instructed learning in the auditory localization pathway of the barn owl. *Nature*, 417, 322–328.

Knudsen EI and Brainard MS (1995). Creating a unified representation of visual and auditory space in the brain. *Annual Review of Neuroscience*, 18, 19–43.

Knudsen EI, Esterly SD, and Lac SD (1991). Stretched and upside-down maps of auditory space in the optic tectum of blind-reared owls: acoustic basis and behavioral correlates. *Journal of Neuroscience*, 11, 1727–1747.

Kornack D and Rakic P (1999). Continuation of neurogenesis in the hippocampus of the adult macaque monkey. *Proceedings of the National Academy of Sciences*, 96, 5768–5773.

Langley P (1987). A general theory of discrimination learning. In D Klahr, P Langley and R Neches, eds. *Production systems models of learning and development*, pp. 99–161. Cambridge, MA: MIT Press.

Lendvai B, Stern EA, Chen B, and Svoboda K (2000). Experience-dependent plasticity of dendritic spines in the developing rat barrel cortex *in vivo*. *Nature*, 404, 876–881.

Leuner B, Falduto J, and Shors TJ (2003). Associative memory formation increases the observation of dendritic spines in the hippocampus. *Journal of Neuroscience*, 23(2), 659–665.

Linkenhoker BA and Knudsen EI (2002). Incremental training increases the plasticity of the auditory space map in adult barn owls. *Nature*, 419, 293–296.

Lipkind D, Nottebohm F, Rado R, and Barnea A (2002). Social change affects the survival of new neurons in the forebrain of adult songbirds. *Behavioral Brain Research*, 133(1), 31–43.

Macnamara J (1982). *Names for things: A study of child language*. MIT Press.

Maletic-Savatic M, Malinow R, and Svoboda K (1999). Rapid dendritic morphogenesis in CA1 hippocampal dendrites induced by synaptic activity. *Science*, 283(5409), 1923–1927.

Martin SJ and Morris RG (2002). New life in an old idea: the synaptic plasticity and memory hypothesis revisited. *Hippocampus*, 12(5), 609–636.

McClelland JL and Jenkins E (1991). Nature, nurture, and connections: Implications of connectionist models for cognitive development. *Architectures for Intelligence: The Twenty-second Symposium on Cognition*, 41–73, Hillsdale, NJ: Erlbaum.

Moser MB, Trommald M, Egeland T, and Andersen P (1997). Spatial training in a complex environment and isolation alter the spine distribution differently in rat CA1 pyramidal cells. *Journal of Comparative Neurology*, 380(3), 373–381.

Newell A (1990). *Unified theories of cognition*. Cambridge, MA, Harvard University Press.

Nottebohm F (1985). Neuronal replacement in adulthood. *Annals of the New York Academy of Sciences*, 457, 143–161.

Nottebohm F (2002). Why are some neurons replaced in adult brain? *Journal of Neuroscience*, 22(3), 624–628.

Nowakowski RS and Hayes NL (2000). New neurons: extraordinary evidence or extraordinary conclusions. *Science*, 288, 771.

Overstreet-Wadiche LS, Bensen AL, and Westbrook GL (2006). Delayed development of adult-generated granule cells in dentate gyrus. *Journal of Neuroscience*, 26(8), 2326–2334.

Paton JA and Nottebohm FN (1984). Neurons generated in the adult brain are recruited into functional circuits. *Science*, 225(4666), 1046–1048.

Piaget J (1970). *Genetic epistemology*. Columbia University Press.

Piaget J (1980). *Adaptation and intelligence: Organic selection and phenocopy*. University of Chicago Press.

Pinker S (1984). *Language learnability and language development*. Harvard University Press.

Poirazi P and Mel BW (2001). Impact of active dendrites and structural plasticity on the memory capacity of neural tissues. *Neuron*, 29, 779–796.

Pouget A, Deffayet C, and Sejnowski TJ (1995). Reinforcement learning predicts the site of plasticity for auditory remapping in the barn owl. In G Tesauro, D Touretzky and T Leen (eds) *Advances in Neural Information Processing Systems 7*, pp. 125–132. MIT Press.

Quartz SR (1993). Neural networks, nativism and the plausibility of constructivism. *Cognition*, 48, 223–242.

Quartz SR and Sejnowski TJ (1997). The neural basis of cognitive development: A constructivist manifesto. *Behavioral and Brain Sciences*, 20, 537–596.

Rakic P (2002). Neurogenesis in the adult primate neocortex: an evaluation of the evidence. *Nature Reviews Neuroscience*, 3, 65–71.

Rakic P (2004). Neuroscience: immigration denied. *Nature*, 427(6976), 685–686.

Ramirez-Amaya V, Escobar ML, Chao V, and Bermudez-Rattoni F (1999). Synaptogenesis of mossy fibers induced by spatial water maze overtraining. *Hippocampus*, 9(6), 631–636.

Rosen DJ, Rumelhart DE, and Knudsen EI (1994). A connectionist model of the owl's sound localization system. In J Cowan, G Tesauro and J Alspector (eds) *Advances in neural information processing systems 6*, pp. 606–614. San Francisco, CA: Morgan-Kauffman.

Rucci M, Tononi G, and Edelman GM (1997). Registration of neural maps through value-dependent learning: modeling the alignment of auditory and visual maps in the barn owl's optic tectum. *Journal of Neuroscience*, 17(1), 334–352.

Sanai N, Tramontin AD, Quinones-Hinojosa A *et al.* (2004). Unique astrocyte ribbon in adult human brain contains neural stem cells but lacks chain migration. *Nature*, 427(6976), 740–744.

Schmidt WC and Shultz TR (1991). *A replication of McClelland's balance scale model.* McGill University, Montreal: McGill Cognitive Science Centre.

Shultz TR (2003). *Computational developmental psychology*. Cambridge, MA: MIT Press.

Shultz TR and Cohen LB (2004). Modeling age differences in infant category learning. *Infancy*, 5, 153–171.

Shultz TR, Mareschal D, and Schmidt WC (1994). Modeling cognitive development on balance scale phenomena. *Machine Learning*, 16, 57–86.

Shultz TR, Schmidt WC, Buckingham D, and Mareschal D (1995). Modeling cognitive development with a generative connectionist algorithm. In TJ Simon and GS Halford, eds. *Developing cognitive competence: New approaches to process modeling*, pp. 205–261. Hillsdale, NJ: Erlbaum.

Siegler RS (1976). Three aspects of cognitive development. *Cognitive Psychology*, 8, 481–520.

Siegler RS (1981). Developmental sequences within and between concepts. *Monographs of the Society for Research in Child Development*, 46, Serial No. 189.

Sirois S and Shultz TR (1999). Learning, development, and nativism: Connectionist implications. *Twenty-first Annual Conference of the Cognitive Science Society*, 689–694, Mahwah, NJ: Erlbaum.

Stanfield BB and Trice JE (1988). Evidence that granule cells generated in the dentate gyrus of adult rats extend axonal projections. *Experimental Brain Research*, 72, 399–406.

Stewart MG and Rusakov DA (1995). Morphological changes associated with stages of memory formation in the chick following passive avoidance training. *Behavioral Brain Research*, 66(1–2), 21–28.

Thivierge JP, Rivest F, and Shultz TR (2003). A dual-phase technique for pruning constructive networks. *IEEE International Joint Conference on Neural Networks*, 559–564.

Valiant LG (1984). A theory of the learnable. *Communications of the ACM*, 27, 1134–1142.

Westermann G (2000). Constructivist neural networks models of cognitive development. *Division of Informatics.* University of Edinburgh, Edinburgh.

Wilkening F (1981). Integrating velocity, time, and distance information: a developmental study. *Cognitive Psychology*, 13(2), 231–247.

Younger BA and Cohen LB (1983). Infant perception of correlations among attributes. *Child Development*, 54(4), 858–867.

Younger BA and Cohen LB (1985). How infants form categories. In G Bower, ed. *The psychology of learning and motivation: Advances in research and theory*, pp. 211–247. New York: Academic Press.

Younger BA and Cohen LB (1986). Developmental change in infants' perception of correlations among attributes. *Child Development*, 57(3), 803–815.

Yuste R and Bonhoeffer T (2001). Morphological changes in dendritic spines associated with long-term synaptic plasticity. *Annual Review of Neuroscience*, 24, 1071–1089.

Zito K and Svoboda K (2002). Activity-dependent synaptogenesis in the adult mammalian cortex. *Neuron*, 35, 1015–1017.

Chapter 5

Modelling cognitive developmental transitions in neural networks: bifurcations in an adaptive resonance theory model

Maartje E. J. Raijmakers

Introduction

Transitions in cognitive development are a major subject in developmental psychology and a central assumption of constructivist theories of development. Computational models, mostly but not exclusively neural networks, have been presented to show that discontinuous transitions in cognitive development can be explained by relatively simple growth principles (e.g., McClelland, 1989; Shultz, Mareschal and Schmidt, 1994; Shultz *et al.*, 1995; van Rijn, van Someren and van der Maas, 2003).

Constructivist principles are central to Piaget's theory of development (e.g., Piaget and Inhelder, 1969). Moreover, Piaget suggested that children pass through developmental stages within which their cognitive abilities remain relatively stable. Transitions between stages correspond to discontinuous changes in problem-solving behaviour and the acquisition of qualitatively new knowledge.

Constructivist theories of cognitive development are involved in two intensely debated issues in developmental psychology: (1) whether discontinuities in behaviour actually do occur, and (2) whether qualitatively new knowledge can ever be acquired. For example, Brainerd (1978) argued that a discontinuity in behaviour is not a testable prediction because it cannot be distinguished from a continuous acceleration in experimental studies. Fodor (1980) provides an example of the second critique, arguing that there are no explanatory mechanistic models of qualitative change.

In the present chapter, we will outline how a developmental transition can be consistently interpreted as a bifurcation—roughly speaking, a bifurcation is a genuine discontinuity in behaviour under gradual change of a parameter; see the Glossary for a more detailed explanation—thereby providing a solution to the two problems mentioned above. However, this interpretation of developmental transitions imposes specific constraints for computational models of transitions. These constraints can be redefined as three criteria to be met by computational models of cognitive developmental transitions. We begin the chapter by describing each of these three criteria for cognitive developmental transitions in detail.

In the second part of the present chapter we discuss the construction of a neural network model of developmental transitions. No fully satisfactory model is presently attainable. However, there are several different ways in which a neural network model that meet the three criteria could be developed. This depends on which aspects of the model are deemed to be most important. Here, we follow a bottom-up approach with a neurological model of connection growth as our starting point. As will be shown, bifurcations appear during development in network models that incorporate neurological growth. Finally, we integrate these results with Adaptive Resonance Theory (ART) (Grossberg, 1976) to show how bifurcations can be involved in the development of representation formation in ART.

Criteria for computational models of cognitive developmental transitions

Computational models of developmental transitions should be consistent with theoretical and empirical insights into cognitive development, which provides some constraints for these models. Interpreting a developmental transition as a bifurcation provides dynamical constraints. Furthermore, a constructivist view of development provides functional constraints. Finally, there is an empirical constraint. The behaviour of the network should match existing empirical data. These constraints will be discussed in turn below. From the constraints, I will derive three criteria for a model of cognitive developmental transitions (henceforth CDT criteria).

Discontinuities modelled by bifurcations

A theory of discontinuities needs testable predictions to distinguish discontinuities from continuous accelerations. Otherwise it can not be decided from empirical data whether development involves genuine discontinuities. Nonlinear dynamical theories, such as catastrophe theory (Thom, 1975), have been

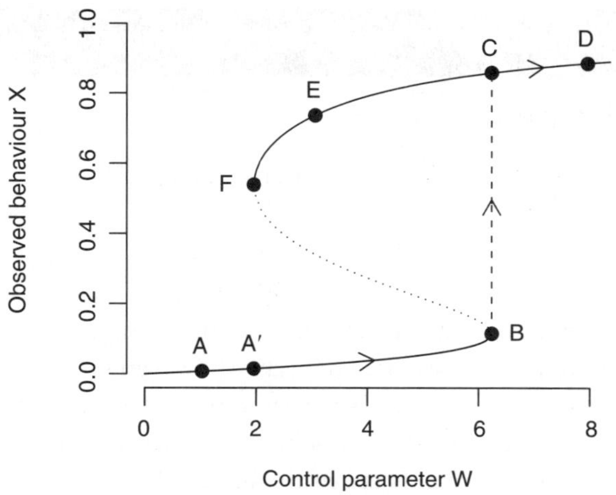

Fig. 5.1 Equilibrium behaviour of a dynamical system with behavioural variable X as a function of parameter W. Solid lines are stable equilibrium states of X, dotted lines are unstable equilibrium states (see text for further explanation).

applied to model discontinuities in cognitive development as bifurcations—also referred to as phase transitions (Preece, 1980; Van Der Maas and Molenaar, 1992). Moreover, catastrophe theory predicts that a system in the vicinity of a transitions bears some specific characteristics, the so called catastrophe flags (Gilmore, 1981). Van Der Maas and Molenaar (1992) formulated the catastrophe flags as testable predictions in empirical studies to distinguish a continuous acceleration from a discontinuous transition. Sudden jump, bimodality of test scores and hysteresis (see Glossary) are three examples of catastrophe flags. A simple form of a bifurcation, and some related flags, will be illustrated in the following example.

Dynamical systems (see Glossary) are defined by equations that describe the evolution of their behaviour over time (see also Kelso, Ding and Schöner, 1992 for a general introduction). If the change in behaviour is a non-linear function of the behaviour itself then bifurcations may occur. As a typical system evolves in time, behaviour first changes, i.e., the behaviour is transient, and subsequently reaches equilibrium, i.e. the behaviour does not change anymore (see Glossary). If a dynamical system in a stable equilibrium state is perturbed, it moves back to the same equilibrium state. If a dynamical system in an unstable equilibrium state is perturbed, it moves away from the unstable equilibrium to another, stable equilibrium state. Bifurcations appear as abrupt changes of the equilibrium behaviour under the gradual change of a parameter, i.e., a control parameter. Figure 5.1 shows the equilibrium behaviour of a typical example, which will be described below. Box 5.1 shows the actual evolution equation

Box 5.1 Example of a nonlinear dynamical system

The evolution equation of X is:

$$\Delta X = \frac{dX}{dt} = -X + (1 - X)Wf(X)$$

$$f(X) = \frac{1}{1 + e^{(\theta - X)/\alpha}}$$

These equations describe the evolution of the average activity, X, of an excitatory network, with average connectivity W (van Ooyen *et al.*, 1995). $\mathrm{d}X/\mathrm{d}t$ is the derivative of X with respect to time, that is, the change of X over time. $f(X)$ is a sigmoid function where α determines the steepness of the function and θ represents the firing threshold. The equilibrium of X given a particular value of W is given by $\mathrm{d}X/\mathrm{d}t = 0$. Figure 5.1 shows the equilibrium values of X with different values of W.

that underlies the temporal evolution of the system. The behavioural variable is X. It depends on the control parameter W. For each value of W, Xhas one or more equilibrium states, as is shown in Figure 5.1. Some equilibrium states are stable (solid line), whereas, some equilibrium states are unstable (dotted line). In Figure 5.1, points F and B are bifurcation points, i.e., fold bifurcation points (Guckenheimer and Holmes, 1993), because at these points the number of equilibrium states changes under variation of the control parameter W. In a fold bifurcation point (which is only one particular kind of bifurcation) a stable and an unstable equilibrium state simultaneously appear or disappear. If the system starts in A and W gradually increases from 1 to 8, then X moves from A to B to C to D. From B to C, the equilibrium value of X makes a sudden jump, which is one of the catastrophe flags. Bimodality of the behavioural variable occurs if W is between 2 and 6. In this area the behaviour of the system is either at the upper plane or at the lower plane. Hence repeated samples of its behaviour will result in a bimodal distribution of X. Bimodality is a necessary, but not a sufficient criterion for a bifurcation.

The flag *hysteresis* (see Glossary) means that the value of W at which a transition takes place depends on the direction of change of W, such that the transition is lagged. If we increase W from 0 to 8, X moves from A, B, C to D, that is X jumps if W $\approx$ 6. If we decrease W from 8 to 0, X moves from D, C, F to A′, that is X jumps if W $\approx$ 2. Hysteresis is a sufficient criterion for the empirical detection of a bifurcation because it is not predicted by other models of rapid changes and discontinuities.

Empirical evidence of developmental transitions in behaviour has been found in several domains, mostly by detecting empirical phenomena associated with the catastrophe flags. Jansen and Van der Maas (2001) present important evidence that the transition from Rule I to Rule II on the balance scale task is a bifurcation. In a series of empirical experiments, they detect several catastrophe flags, such as bimodality and hysteresis, in children's performance. For example, sudden jumps are detected in conservation of liquid experiments (Raijmakers, Van Koten and Molenaar, 1996). However, bimodality of the behavioural variable and sudden jumps in the behavioural variable do not occur in McClelland's (1989) connectionist Parallel Distributed Processing (PDP) model of learning the balance scale task (Raijmakers, *et al.*, 1996). Empirical evidence for bifurcations has also been found in the domain of language development (e.g., Dromi, 1987), motor development (e.g., Wimmers, *et al.*, 1998), and analogical reasoning (Hosenfeld, *et al.*, 1997).

Qualitatively new knowledge by self-organization

The key idea that underlies the notion of developmental transitions is that qualitatively new knowledge is acquired in the presence of a transition. Several researchers (remarkably also Fodor, 1981: 300) suggested that the occurrence of a bifurcation (often mentioned as self-organization) might be a mechanism by which a non-linear dynamical system, such as a neural network, acquires qualitatively new knowledge (cf. Bereiter, 1985; Lawson and Staver, 1989; Molenaar, 1986; Molenaar and Raijmakers, 2000). The impossibility of acquiring qualitatively new knowledge is still debated (e.g., Laurence and Margolis, 2002), often in response to Fodor (1980, 1981). According to Fodor, learning is necessarily some sort of inductive inference involving hypotheses formation and confirmation. Therefore concepts that a subject can learn have to be composed of innate primitive concepts. New concepts that cannot be composed of innate primitive concepts are not included in the hypothesis space of a subject. These concepts are qualitatively new and cannot be learned—hence a system can never become more powerful by learning. The addition of conceptual resources (by maturation) is the only way in which Fodor envisages a system to become more powerful. In accordance with the literature, we will refer to this as Fodor's learning paradox (see also Shultz, Mysore, and Quartz, Chapter 4 this volume).

Connectionist models of learning and cognitive development have been proposed as systems that can acquire new concepts (e.g., Elman *et al.*, 1996). After learning input—output relations, the hidden node activity in connectionist models is considered to represent qualitatively new concepts. An objection against this claim is that the learned representations thus obtained still have to be compositional in the sense of Fodor's learning paradox (e.g. Bloom and Wynn, 1994).

In contrast to neural networks with a fixed number of hidden nodes, cascade correlation networks (Fahlman and Lebiere, 1990) model constructive growth principles by the addition of hidden nodes as a function of learning (see Quinlan, 1998, for alternative architectures). Quartz (1993) argues that these networks become more powerful because they increase their hypothesis space during learning, and hence escape Fodor's learning paradox (see also Mareschal and Shultz, 1996). Cascade-correlation networks are presented as models of transitions in development on several Piagetian task domains: the balance scale (Shultz *et al.*, 1994), distance, space and velocity (Buckingham and Shultz, 2000), and seriation (Mareschal and Shultz, 1999; see also Shultz *et al.*, Chapter 4 this volume).

However, the constructive growth principles in generative neural networks would not seem to escape Fodor's learning paradox. In fact, far from disproving Fodor's critique, generative neural networks constitute a demonstration of the validity of this critique (Raijmakers, 1997; Marcus, 1998). During learning, a generative neural network is extended with new hidden nodes. Consequently, the input–output function of the network is extended with new elements, which increase the representational power of the network. Moreover, these added elements are not equivalent to existing elements because the connection structure of each added element is new. Fodor (1980) argues that the power of a system cannot be increased unless conceptual resources are added. In a generative neural network, a hidden node should be considered as such a resource. Consequently, these generative neural networks do not provide a genuine alternative to Fodor's point of view.

The occurrence of a bifurcation may be a real alternative to the addition of resources as a way to escape Fodor's learning paradox. If the behaviour of the example system in point D (Figure 5.1 and Box 5.1) differs qualitatively from that in point A, then the system becomes more powerful without adding resources. In non-linear dynamical systems, a qualitative change of the dynamics is triggered by a gradual change of a parameter. The change of a parameter is not an addition of resources. The parameter is a resource that was present all the time. Moreover, the change of the parameter does not by itself determine the nature of the qualitative change, that is, the increase of power. Thus, the challenge is to define a neural network which shows a bifurcation related to development. Moreover, the occurrence of the bifurcation should be associated with an increase in the power of cognitive behaviour.

Rule-like behaviour

In the study of cognitive development, transitions are related to learning new rules or strategies. Neural network models of transitions in cognitive development have been applied to learning in proportional reasoning tasks, such as the balance scale task and the distance, space and velocity task. The most important empirical finding is that children learn these tasks by applying increasingly complex rules. Even children as young as five years have been found to perform in a rule-governed fashion (Siegler, 1978).

Siegler (1981) designed a methodology to assess rules from behavioural data. The central idea is to compare observed patterns of responses to different problem types with theoretically expected response patterns. Expected response patterns are derived from rules that children are expected to apply. Observed response patterns are attributed to a rule according to the best matching rule pattern. However, in some cases this method is too tolerant because it lacks statistical tests to decide whether the rule patterns are significant, latent structures of the data. Latent class analysis provides a statistically reliable way to assess rules in categorical data (Jansen and van der Maas, 1997). Moreover, in contrast to the rule-assessment methodology, latent class analysis can be applied both in a confirmatory and exploratory fashion. Hence, using this method, the following questions can be tested reliably: Which of the expected rules are needed to describe the data? Are rules applied at all? Are other rules applied than expected?

Latent class analysis of empirical data confirms the application of rules by children and adults (Jansen and Van der Maas, 1997). In contrast, latent class analysis of response patterns generated by McClelland's (1989) model finds no evidence of rule-like behaviour by the neural network (see also Raijmakers, van Koten and Molenaar, 1996). This shows that a statistically reliable method for detecting rules is important, especially in case justifications of responses are unfeasible. The behaviour of the cascade-correlation (CC) model of the balance scale task (Shultz *et al.*, 1994) has also been subjected to latent class analysis. Quinlan *et al.* (in press) found that the CC models do recover certain rules of operation similar to those of humans. However, use of the torque rule (Rule IV) is not found, and the networks also acquire rules never previously seen in humans.

When during the learning of a regular behaviour, such as rule IV of the balance scale task, an individual shows a sequence of increasingly complex rules instead of an incremental improvement of the target behaviour, the underlying knowledge requires concept-based, selective representations in contrast to item-based, non-selective representations (Kendler, 1995). Note that the rules may still be perceptual rules as are found, for example, in perceptual classification tasks (Thompson, 1994). Representations in neural networks are mostly analysed by looking at the hidden nodes. Factor analyses, or cluster analyses of hidden unit activations are used to show that concepts are learned and are represented by a distributed code (as in Elman *et al.*, 1996). However, to decide whether representations are concept-based (selective) or item-based (non-selective), one should also look at the overt behaviour of the network. The discrimination-shift paradigm (Kendler, 1995), which is extensively used with infra-humans, children and adults, is particularly suited to test neural networks. Raijmakers *et al.* (1996) showed that in a feedforward PDP-model resembling McClelland's (1989) model of the balance-scale task, the representations learned while performing a discrimination-shift tasks, resemble those of infra-humans, and not those of human adults or children. Nevertheless, an analysis of the hidden node representations suggests that the network learned concepts, that is, selective representations. In contrast, CC networks (Sirois and Shultz, 1998) without a hidden layer can meet the criteria for forming nonselective representations, as do children. Furthermore, if the CC networks include some rehearsal time that is, an increased amount of training during each stimulus presentation, the network meets the empirical criteria for forming selective representations, as do adults.

The outcomes of the empirical studies with neural networks underscore the importance of analysing both the network's overt behaviour as well as applying rigorous statistical techniques when deciding whether a network has acquired rules and concept-based representations.

Conclusion: three criteria for a model of cognitive developmental transitions

Neural network models of transitions in cognitive development have multiple goals. As a general goal, a neural network model of cognitive development can show how low-level change mechanisms explain high-level cognitive changes. In the ideal case, there is a clear relation between the change mechanisms of the model and neurological properties of the brain. In that case, a neural network model shows a possible relation between neurological mechanisms and cognitive phenomena.

With regard to modelling *cognitive developmental* transitions, a more basic question needs to be answered: "Can qualitatively new knowledge be acquired?" Because a consistent interpretation of transitions in development is still under discussion, a neural network model could be an existence proof of the acquisition of qualitatively new knowledge. The first goal requires a good agreement between empirical data and network behaviour. The second goal requires that the model agrees with a consistent, theoretical interpretation of developmental transitions. In the foregoing section, we discussed the theoretical and empirical context of developmental transitions that we consider to be relevant to computational models. We can formulate the associated constraints as three criteria that a fully satisfactory model of cognitive developmental transitions. We call these criteria cognitive developmental transitions criteria, i.e., CDT criteria.

The first criterion is that the model shows a bifurcation to explain a transition in cognitive development. Bifurcations can be detected by the mathematical and numerical methods of bifurcation analysis (Guckenheimer and Holmes, 1993; Borisyuk and Kirillov, 1992) and the empirical methods associated with detecting catastrophe flags. The second criterion is that the behaviour of the model after the bifurcation is qualitative new compared to the behaviour before the bifurcation. Quartz (1993) proposes a formal measure to define qualitative differences in the behaviour of a learning system. The third criterion is that the model shows rule-like behaviour and learns conceptual representations. Reliable statistical techniques are needed to demonstrate rule-like behaviour. Learning conceptual representations can be demonstrated by empirical methods, such as the discrimination-shift paradigm.

I believe that the current status of neural network models of developmental transitions is such that a focus on only one or two of the CDT criteria may already be sufficient to provide important insights in how to develop a complete model. In the remaining part of this chapter we will discuss a model that focuses on the first criterion, showing a bifurcation that is related to development and which affects cognitive behaviour. In doing this we start from a neurological model of connection growth (Van Ooyen *et al.*, 1995) that shows a bifurcation during development. Then, we integrate this bifurcation with Adaptive Resonance Theory (ART) (Grossberg, 1976) to show how a bifurcation can be involved in the development of representation formation in ART.

A neurocomputational model of transitions

Bifurcations in a neurological model of connection growth

Biological models of neurite outgrowth show that constructive growth principles can be implemented by the evolution of a fixed number of variables. Van Ooyen *et al.*'s (1995) model is called a constructivist model because the development of network morphology is directly dependent on neuron activity and external influences upon it. The mathematical basis of the model consists of evolution equations for neuronal activity of individual neurons and the outgrowth of the neuritic fields of individual neurons, which are mutually dependent (see Box 5.2). The network is a so-called shunting neural network (SNN) (Grossberg, 1988). In an SNN, excitatory inputs drive the membrane potential towards a finite maximum, while inhibitory inputs drive the membrane potential towards a finite minimum. In the model discussed below, the neurons are placed on a circle such that connections grow first between neighbors and then with more distant neurons.

Box 5.2 A neurological model of connection growth

The neuronal activity of the shunting neural network is described by membrane equations, which are transformed into the following (dimensionless) set of equations for excitatory activity x_i $(1 \leq i \leq \mathrm{N})$ and inhibitory activity y_j $(1 \leq j \leq \mathrm{M})$.

$$\frac{dX_i}{dT} = -X_i + (1 - X_i)\sum_k^N W_{ik} f(X_k) - (H + X_i)\sum_l^M W_{il} f(Y_l)$$

$$\frac{dY_j}{dT} = -Y_j + (1 - Y_j)\sum_k^N W_{jk} f(X_k) - (H + Y_j)\sum_l^M W_{jl} f(Y_l)$$

$f(x)$ is the firing rate function described in Box 5.1. X_i is the mean membrane potential of excitatory cell i, Y_j is the mean membrane potential for inhibitory cell j. All potentials are normalized between 0 and 1. W_{ik} is the connectivity from neuron k to i. $H = A/-B$, where A and $-B$ are the saturation potentials of excitatory and inhibitory cells respectively. $T = t/\tau$, where τ is the membrane time constant.

The outgrowth of individual neurons, R_i, is dependent upon the firing rate:

$$\frac{dR_i}{dT} = \rho G(f(X_i))$$

$$G(f(X_i)) = 1 - \frac{2}{1 + e^{(\varepsilon - f(x_i))/\beta}}$$

where ε is the value of $f(X_i)$ for which $G = 0$ and β determines the steepness of the function. Depending on $f(X_i)$, a neuritic field will grow out ($G > 0$ when $f(X_i) < \varepsilon$), retract ($G > 0$ when $f(X_i) > \varepsilon$), or remain constant ($G = 0$ when $f(X_i) = \varepsilon$). For more details see Van Ooyen *et al.* (1995).

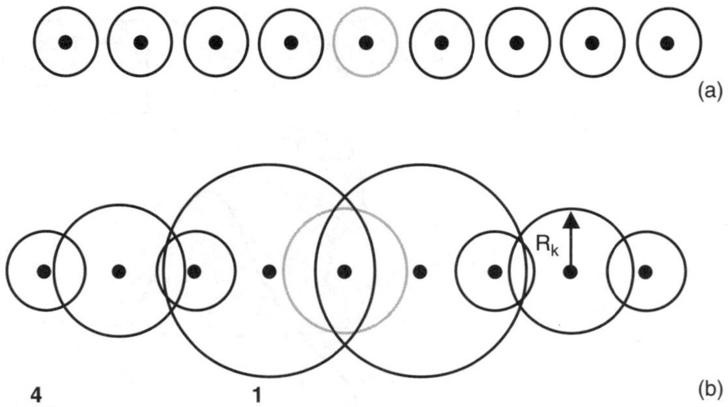

Fig. 5.2 A schematic view of a one-dimensional layer of eight excitatory (black lines) and one inhibitory (grey line) neurons. (a) the model starts with non-overlapping neuritic fields, that is disconnected neurons. (b) In equilibrium, a pattern of small and large overlapping neuritic fields has resulted, which means that neurons are connected with different weights. Replication of results in Van Ooyen *et al.* (1995).

Figure 5.2 shows a simple example of the model. The neuritic field of each neuron, which is shown by a circle, is the range of its interactions with other neurons. The neuritic field of neuron k is modelled by one real-valued parameter, its radius, R_k. The connection weight between two neurons k and l equals the overlap of their neuritic fields.

In the model, initially disconnected cells (Figure 5.2a) organize themselves into a network of connected cells (Figure 5.2b) under the influence of intrinsic activity. Several characteristic phenomena of brain development emerge from the dynamics of this relatively simple model, such as a transient overproduction (overshoot) in the number of connections (see van Ooyen *et al.*, 1995). The bifurcation diagram of Figure 5.1 shows the dependence of the average activity, X, on the average connection weight, W. Due to their mutual dependence, in time, X and W move approximately from A to B to C to E. The effect of the bifurcation is illustrated in Figure 5.3, which shows that the average activity of excitatory neurons abruptly increases with gradual change of connection weights.

The dynamics of the model are defined at a low level, that is, at the level of the interaction between individual cells. Morphological and functional properties of the model, such as recovery of network structure after cell death and a critical period for the elimination of neurons, appear as emergent properties of the system (Van Ooyen *et al.*, 2003, 1995, 2003).

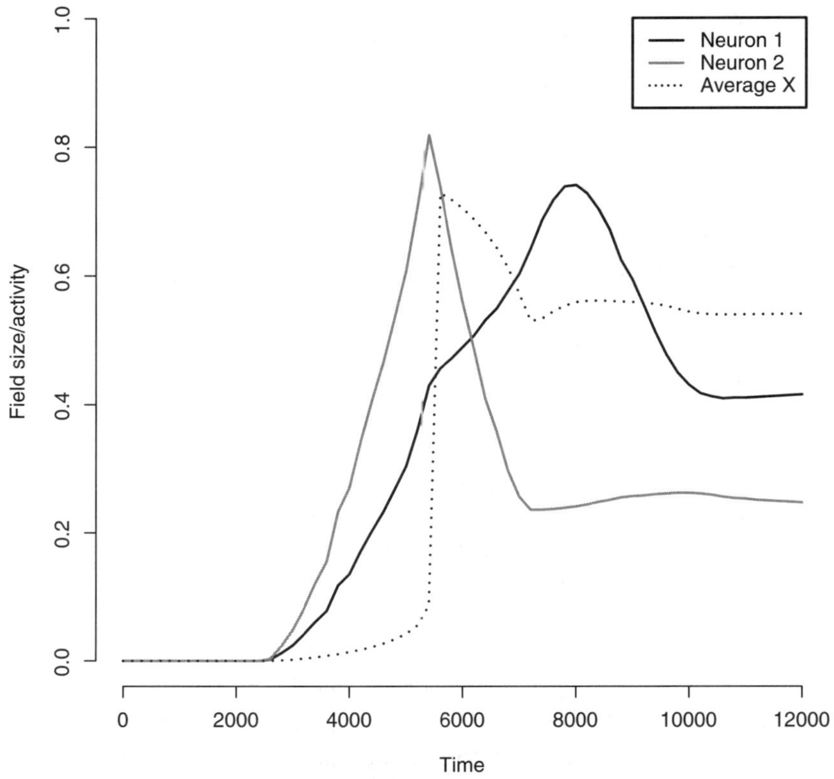

Fig. 5.3 The time course of the neuritic field size of two neurons, 1 and 4, and of the average activity of excitatory neurons of the model displayed in Figure 5.2. Note the initial overshoot of the neuritic field sizes of the neurons and the abrupt change of the average activity. Replication of results in Van Ooyen *et al.* (1995).

Bifurcations in adaptive resonance theory

The model of morphological growth described above, does not, as such, show functional, cognitive behaviour (e.g., CDT criterion 2). Yet, in some neural network models of cognition, such as ART (Grossberg, 1976), the SNN implements a content addressable memory (see Glossary). In ART, the SNN maps input patterns to category information. The category information is stored in long-term memory, which is recalled directly on the basis of the content of the input. We will show how the occurrence of bifurcations in the SNN leads to qualitative changes of ART at the cognitive level. First, we will briefly introduce an ART neural network for unsupervised classification.

Adaptive Resonance Theory

ART defines network principles for unsupervised learning of patterns. As in a statistical clustering algorithm, patterns that are alike according to some distance measure, are clustered into one category. In ART, each category is represented by a vector, which describes the common factors of its elements. An ART network constructs categories online during the presentation of input vectors, and is able to classify known and unknown input vectors. Each sequence of input vectors will result in a stable clustering. Moreover, an ART network remains adaptable to unknown input vectors and a very long or frequent presentation of one pattern will not remove some of its learned categories. There exist many implementations of ART: ART 1 (Carpenter and Grossberg, 1987a) is a classifier of binary vectors; ART 2, ART 3 and fuzzy ART are classifiers of analog vectors (Carpenter and Grossberg, 1987b, 1991; Carpenter, Grossberg and Rosen, 1991). Unsupervised ART networks are extendable to ARTMAP networks for supervised learning (Carpenter *et al.*, 1998). See Levine (2000) for an overview of ART architectures. Here we will describe Exact ART (Raijmakers and Molenaar, 1997), because it actually implements the content addressable memory as an SNN. Exact ART clusters analog input vectors.

The Exact ART model consists of several modules (Figure 5.4): an input level (F0), a feature level (F1), a category level (F2), and a reset system. Between F0 and F1 only bottom-up, non-adaptable connections exist. F1 and F2 are fully connected by bottom-up, adaptable connections (in Figure 5.4 shown as lines with filled circles near F2) and top-down, adaptable connections (in Figure 5.4 shown as lines with filled circles near F1). The adaptable connections form the long-term memory (LTM) where the learned category information is stored. The reset system receives input from F0 and F1 and its output is directed to F1 and F2.

ART is based on the idea of adaptive resonant feedback between two layers of nodes (here F1 and F2) as developed in Grossberg (1976). Briefly, the learning and classification of input patterns goes as follows. If the network receives input, the best matching learned category is searched by a partly parallel, partly serial process. A matching category results in resonant feedback between F1 and F2. Resonant feedback is sustained activity in the system, representing both the input (in F1) and the matched category (in F2). This means that the input pattern is stored in short-term memory (STM). Storage in STM results in the adaptation of the matched category by changing the relevant adaptive weights, that is, storage in LTM. In detail, the following sequence of events occurs after presentation of an input pattern. Steps 1 to 9 correspond with diagrams 1 to 9 in Figure 5.5.

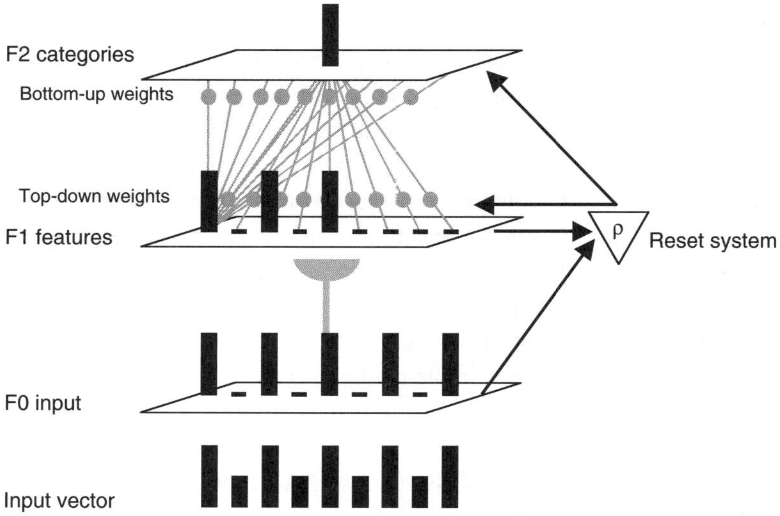

Fig. 5.4 General layout of Exact ART. Black bars are activities at the different levels, grey lines are connections between levels. Between F0 and F1 only bottom-up, non-adaptable connections exist. F1 and F2 are mutually, fully connected by bottom-up connections (filled circles near F2) and top-down connections (filled circles near F1). Only outgoing connections of one F1-node and one F2-node are drawn.

Step 0 Initialize adaptable weights between F1 and F2. (Top-down weights are initialized to 0, bottom-up weights are initialized to small random values.) Set ρ, $0 \leq \rho \leq 1$. The fraction ρ is the vigilance threshold which indicates how close an input must be by a learned category to match.

Step 1 In F0, the input vector is preprocessed, such that contrast is enhanced, activity is normalized, and noise is suppressed.

Step 2 The resulting activity vector is used as bottom-up input for the feature level, F1. Without the influence of learned categories, the activity vector of F1 equals the activity vector of F0.

Step 3 The activity vector in F1 is multiplied by bottom-up, adaptable weights and triggers an activity pattern at the category level, F2. In its standard form, F2 is an SNN that results in a winner-takes-all competition between F2-nodes (equations of the SNN in F2 are given in Box 5.3). This means that the F2-node with the largest input is the only node with sustained activity. The one, active F2-node is the recognized category. The category is represented by its connected bottom-up weight vector (In Figure 5.3, the thickness of the line denotes the size of the weight). In a standard F2-layer, categories have

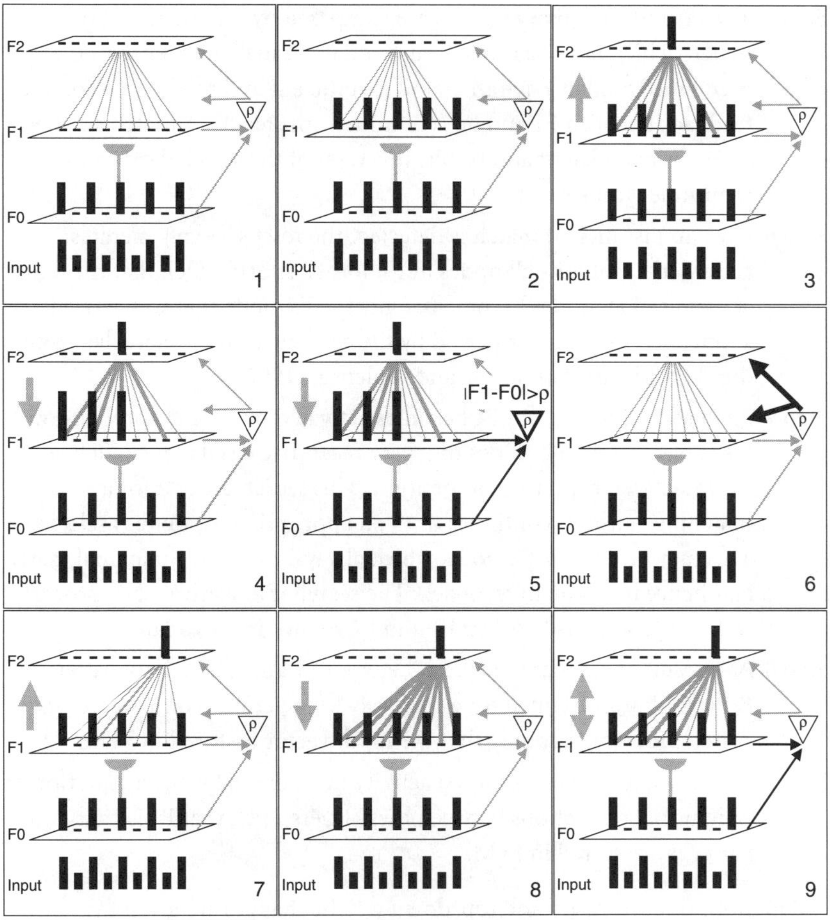

Fig. 5.5 Each diagram shows a phase of the classification and learning process of Exact ART. Separate steps are described in the text.

a local representation, because they are represented by the weight vector of one F2-node.

Step 4 The active F2-node also sends activity down, which generates the expected activity pattern according to the active category. This top-down input for F1 is the F2 activity vector multiplied by top-down weights. F1 now receives two input vectors: the bottom-up input from F0 and the top-down expectation from F2. The resulting F1 activity consists of the matching elements of both vectors. Resulting F1 activity is normalized such that the sum of F1-node activities is 1.

Step 5 The F0 and F1 activity vectors are compared by calculation of the match, which is a scalar between 1 and 0[1]. A match larger than the predefined vigilance parameter ρs is sufficient and results in resonant feedback between F1 and F2 (continue with step 8). A match smaller than ρ is insufficient and results in a reset of the selected category (continue with step 6).

Step 6 Now an insufficient match is detected, the reset system generates activity that inhibits all nodes of F1 and F2 shortly, such that the activities of all nodes become 0. Only the F2-node that was active is suppressed for a longer period by the construction of a so-called gated dipole field (see Raijmakers and Molenaar, 1997).

Step 7 As in steps 2 to 5, again F1 becomes activated by F0 and competition takes place in F2. F2-nodes that very recently caused a mismatch, are now excluded from the competition. This search process for a matching category continues until an appropriate F2 node is selected. This may be a blank F2-node, which always results in a maximal match and hence in resonant feedback. The search is a relatively fast process, so that the weights between F1 and F2 get no time to adapt.

Step 8 As a result of resonant feedback between F1 and F2, activity patterns in F1 and F2 are sustained for a relatively long period (compared to the search process), meaning that the input vector is stored in STM.

Step 9 The consequence of sustained activity in F1 and F2 is the adaptation of weights of connections between these layers, such that F1 activity patterns are stored in LTM.

Equations of bottom-up and top-down weight changes are given by:

$$\begin{aligned} Top\text{—}down : \Delta z_{ji} &= \alpha f(y_j)(f(x_i) - z_{ji}) \\ Bottom\text{—}up : \Delta z_{ji} &= \alpha f(y_j)(f(x_i) - z_{ji}) \end{aligned} \tag{1}$$

Δz_{ji} denotes the change in weight from F2-unit j to F1-unit i. Δz_{ij} denotes the change in weight from F1-unit i to F2-unit j. y_j denotes the activity of F2-node j, x_i denotes the activity of F1-node i, f() is a threshold function. α denotes the learning rate. Bottom-up weights are initialized at small random values, top-down weights are initialized at zero. According to Equation 1, weights are only updated if they are connected to active F2-units. Hence, learned categories are not affected by presentation of unrelated input patterns, which prevents catastrophic forgetting. Moreover, weights are limited in size, because F1 activity is normalized and during the update weight z_{ji} and z_{ij} become equal to x_i. Although from the above description, the classification

Box 5.3 Shunting neural network

The shunting neural network in the F2 layer of Exact ART strongly resembles the shunting neural network as described in Box 5.2. However, here the connection structure is on-centre off-surround, which means that each node is excitatory connected to itself (on-centre) and inhibitory to all other nodes (off-surround). This is described by the following equation, which forms the main structure of the F2-layer.

$$\frac{dy_j}{dT} = -y_j + (1 - y_j)\left(h(y_j) + \sum_{k}^{N} f(z_{kj}x_k)\right) - y_j \sum_{l \neq j}^{M} h(y_t)$$

y_j is the activity of node j in F2 ($1 \leq j \leq M$). x_k is the activity of node k in F1. $h()$ and $f()$ are non-linear functions. z_{kj} is the bottom-up adaptive weight from node k in F1 to node j in F2. The first term $h(y_j)$ corresponds to the self-excitation of F2-node j. The first term corresponds to the activity delay. The second term corresponds to the excitatory input. This consist of the self-excitation, $h(y_j)$, and the F1 input for F2-node j. The last term corresponds to the inhibitory influence of all F2-nodes except for node j.

1. The match, R, is calculated as follows: $R = 1 - \frac{1}{2}\sum_{i=1}^{N} |u_i - x_i|$. u_i is element i of the activity vector in F0 and x_i is element i of the activity vector in F1. Both vectors have length N.
2. ART3 and Distributed ARTMAP can also form distributed code (Carpenter and Grossberg, 1991; Carpenter *et al.*, 1998). As is pointed out by Carpenter (2001), if distributed codes are learned, there is a trade-off between adaptability and stability of learned categories, which does not occur in local coding.

and learning process appear to be sequential and step-wise, the actual implementation is continuous and parallel in Exact ART (but not in most other ART implementations).

From local to distributed representations

In its standard form, the category level of ART models is implemented by winner-takes-all dynamics in an SNN (see Box 5.3), which means that only one F2-node can be active simultaneously. As a consequence, the representations of categories are local: each category is represented by one weight vector and each weight is part of the representation of only one category. However, like in the biological model of neurite outgrowth, the SNN appears to have several dynamic regimes, depending on the range of connections in the network. Raijmakers *et al.* (1997) performed a numerical bifurcation analysis to systematically map all relevant bifurcations and dynamical regimes of the SNN as it is implemented in Exact ART. To this end, they varied structural parameters of the SNN. That is, they varied the range of the excitatory and inhibitory connections. This was accomplished by setting the inhibitory and excitatory connection weights within F2 according to a Gaussian function with parameters σ (as the standard deviation in a normal distribution function) and d (the surface under the graph, which equals 1 in a normal distribution function). σ_i and $\sigma_e$denote the range of inhibitory and excitatory connections respectively; d_e and d_i denote (per neuron) the sum of connection weights of, respectively, excitatory and inhibitory connections (See Figure 5.6).

SNN's with on-centre and off-surround connections are found in several neural structures, e.g. in the hippocampus, the neocortex, and the cerebellum (Anderson *et al.*, 1969; Stefanis, 1969; Eccles *et al.*, 1967; Kandall, Schwartz

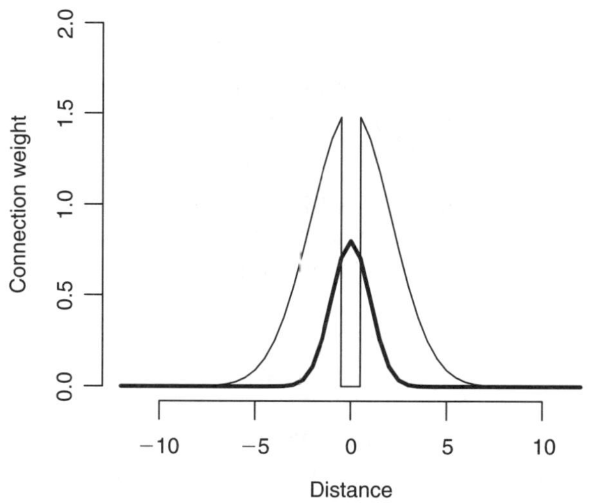

Fig. 5.6 Graphs of the strength of excitatory (bold line) and inhibitory (thin line) connections between units within F2. The horizontal axis represents the distance between two neurons. The vertical axis represents the connection strength of two neurons. Note that there are no self-inhibitory connections, which means that the connection weight of inhibitory connections with zero distance equals zero. For these graphs the following parameter values are used: $(\sigma_e, \sigma_i, d_e, d_i) = (1, 2.1, 1, 6.48)$.

and Jessell, 1991). During postnatal maturation several structural properties of these networks change. The main properties of cortical networks that change are the length of axons and dendrites, the number of spines, the number of synaptic connections, myelination of axons, the number of glial cells, and the vascular tissue (Purves, 1994). Parameters σ_e, σ_i, d_e and d_i can be interpreted as properties of cortical networks that change during postnatal development. σ_e and σ_i represent the length of axons and dendrites and the density of synapses and spines are represented by parameters d_e and d_i. The other properties of cortical networks that change during postnatal development have no equivalent in the SNN.

Numerical bifurcation analysis shows that fold bifurcations (see Figure 5.1) appear in the networks' activity dynamics, with σ_i and σ_e as control parameters. Figure 5.7 shows different activity patterns of a 25-nodes SNN as a function of σ_i.

The most striking difference between the activity patterns in Figure 5.7 is the number of simultaneously active F2-nodes. The winner-takes-all dynamics corresponds with one active node (as with σ_i equals 8.15 in Figure 5.7) and results in local representations. If more nodes are simultaneously active, that is, with lower values of σ_i, distributed representations may also appear. This is tested in the following, small simulation study. In this simulation, Exact ART (with fixed F2 connections) learns to classify a small number of input patterns. The study is constructed after Carpenter and Grossberg (1987b).

Exact ART is repeatedly presented the following sequence of input patterns A B C A D, which are shown in Figure 5.8. The match criterion, ρ, is set to 0.7 such that Exact ART with a winner-takes-all dynamics would learn four distinct categories to represent the input patterns. The total number of F2 nodes is 25. In this study σ_i is relatively small (σ_i equals 1.3), such that 7 nodes are active simultaneously. Consequently, the total number of nodes in F2 is too small to form four totally distinct representations ($4 \times 7 > 25$).

After presenting this sequence of inputs several times (after 10 presentations no important changes take place) all inputs are classified in different categories. Although each input pattern is represented by a unique set of F2-nodes, some F2-nodes are active during the presentation of several input patterns. Nodes that are active during the presentation of several input patterns develop another function than nodes that are only active during the presentation of one input pattern. Hence, distributed representations are learned. The representations are stable and the match between input, i.e. F0, and learned representation, i.e. F1, is sufficient for each input pattern, although it is no

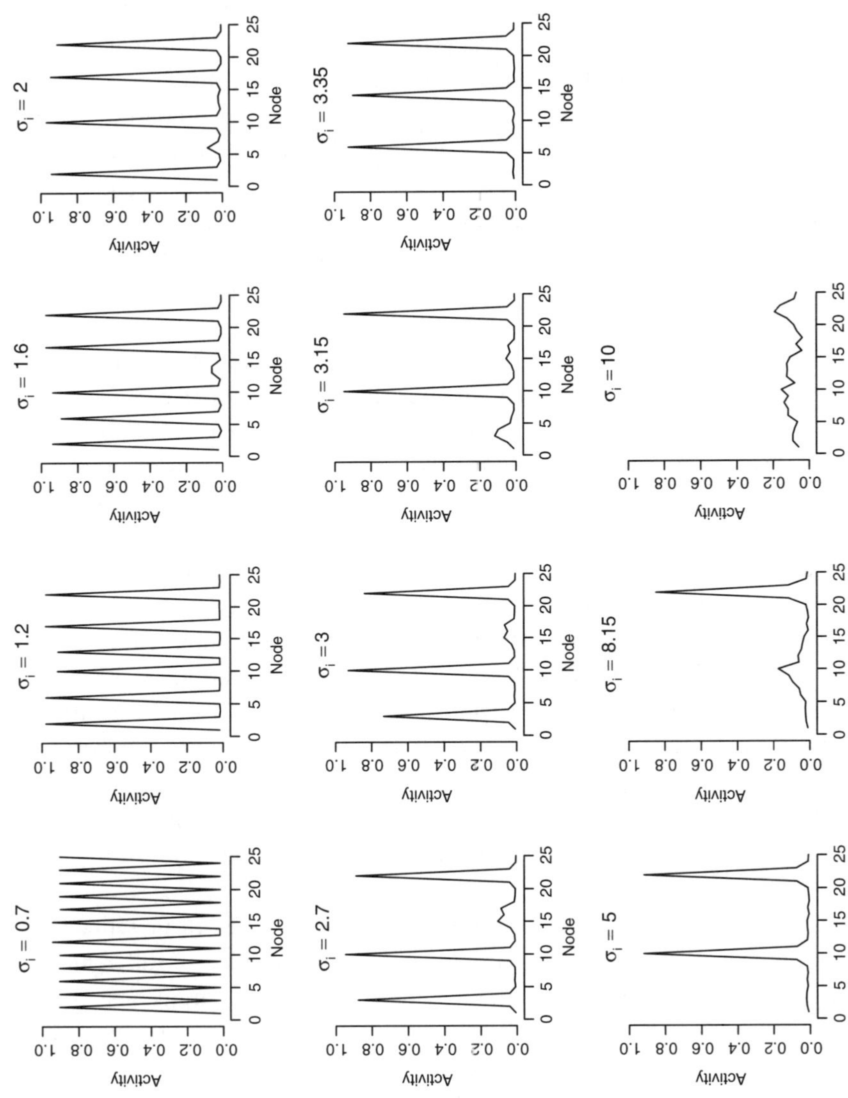

Fig. 5.7 Activity patterns for different values of σ_i, ranging from 0.7 to 10.0 ($d_e = 1$, $d_i = 24$, $\sigma_e = 1.02$). In general, the number of active nodes decreases with increasing σ_i: i.e., 12, 6, 5, 4, 3, 3, 2, 3, 2, 1, and 0, with some exceptions. More details can be found in Raijmakers *et al.* (1997).

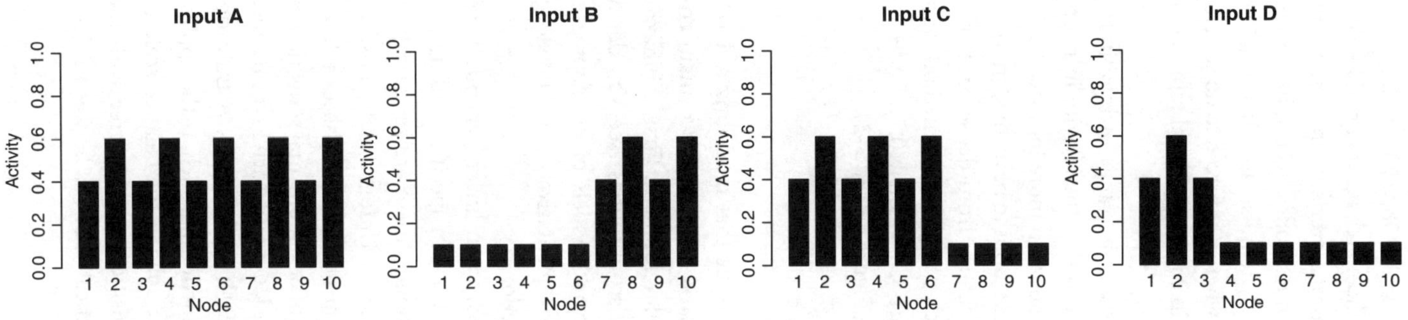

Fig. 5.8 Input patterns A, B, C, and D of a small simulation study (see also Carpenter and Grossberg, 1987b). Horizontal axis denotes node index, vertical axis denotes the size of activity.

longer perfect; The match is 0.99, 0.98, 0.73, and 0.81 for inputs A, B, C, and D respectively (calculation of the match is explained in Note 1). As expected, the learned category representations overlap: representations of input patterns C and D share five out of seven F2-nodes. As can be verified in Figure 5.8, of all pairs, input patterns C and D are most alike. It appears that Exact ART learns distributed representations where it is forced to do so. These distributed representations are unique, but, in contrast to local representations, they represent similarities between patterns.[2]

During development, overall, the range of inhibitory connections, that is σ_i, increases. This would mean that the activity in neuronal networks in the cortex changes from more global to more localized, undergoing one or more bifurcations. For the representations to be formed in ART networks, this would mean that representations change from dense codes to sparse codes. Dense coding would mean that each state is represented on average by about half the units being active (e.g., Figure 5.7, σ_i equals 0.7). Codes with low activity ratios are called sparse codes (e.g., Figure 5.7, $\sigma_i > 2$).

Conclusion

In the present chapter, we argue that modelling a developmental transition by a bifurcation (see Glossary) may solve two main criticisms of constructivist theories of cognitive development. First, a bifurcation model provides testable criteria to detect discontinuities (see Van der Maas and Molenaar, 1992). Second, neural network models of developmental mechanisms that show bifurcations, may learn qualitatively new knowledge. That is, they may escape Fodor's learning paradox.

Neural networks are usually non-linear dynamical systems (see Glossary) and they often show bifurcations in specific conditions (see for example van der Maas *et al.*, 1990), although these are rarely explored in psychological models (but see Rodriguez, 2001). Unfortunately, most current neural network models of developmental transitions do not show relevant bifurcations. Current neural network models of developmental transitions are usually applied to a cognitive developmental task domain, such as the balance-scale task (McClelland, 1989; Shultz *et al.*, 1994; Roth, 2000). A transition in these domains involves higher cognitive functioning, such as the application of increasingly complex, cognitive rules for solving the balance scale task. Neural network models of this type of transition should agree with the empirical data on learning rules. More theoretical relations between rules and bifurcations are described in van der Maas (1998).

These observations led to the formulation of the three CDT criteria (cognitive developmental transitions criteria) that, in our opinion, neural network models of cognitive developmental transitions should obey:

1. The model should show a bifurcation to explain transitions in cognitive development.
2. The behaviour of the model after the bifurcation should be more powerful than before the bifurcation.
3. The model should show rule-like behaviour and should learn conceptual representations.

To our knowledge, there is currently no model that obeys all three CDT criteria. Hence, the CDT criteria should be viewed as guidelines to evolve models of cognitive developmental transitions. In the present chapter, we discussed one such model in which bifurcations occur that are related to development. These bifurcations are derived from a neurological model of connection growth in the cortex. We examined how the bifurcations in this model are related to functional, cognitive behaviour. The model is an ART (Exact ART) model for free classification of input patterns.

In the Exact ART model, bifurcations appear in the activity dynamics of its content addressable memory (i.e., the category layer). The control parameter of these bifurcations is the (fixed) connection structure within the category layer, in particular the range of inhibitory connections. The bifurcations induce the learning of different representational regimes in Exact ART, ranging from local to distributed representations.

Overall, during development, cortical connections increase in range. A temporary overshoot of cortical connection range, which implicates also a decrease in connection range, was identified in the literature (e.g., Uylings *et al.*, 1990) and modelled by Van Ooyen *et al.* (1995). Yet the overall developmental trend is for an increase of connection range. This would imply that during development the parameters that we studied in the Exact ART model generally increase. Consequently, what the model predicts is a developmental change in the representational regime from dense coding to sparse coding. It is likely that sparse coding is used in adult cortical networks (Földiák and Young, 1995; Young and Yamane, 1992). However, not so much is known about activity patterns in immature cortical networks as a consequence of external input. Van Pelt *et al.* (2004) study spontaneous activity patterns in cultured, growing neural networks. They report a subsequent gradual increase and drastic decrease of burst intensity. However, it remains to be seen whether this situation is comparable to externally induced activity in cortical networks *in vivo*.

In general, the representational regime of a neural network induces functional constraints on the network (i.e., functional brain constraints). Földiák and Young (1995) point out the functional differences between dense and sparse coding schemes in networks. Memory capacity is highest for sparse codes, limited for local codes and low for dense codes. Speed of learning is very fast for local codes, fast for sparse codes and slow for dense codes. Generalization is good for sparse and dense codes, but none for local codes. Interference is none for local codes, controlled in sparse codes and strong in dense codes. Finally, simultaneous items can be active in an unlimited way in local codes, several items can be active in sparse codes, and only one in dense codes.

Functional changes due to a bifurcation are relevant to the second criterion. A change from dense to sparse coding would imply an improvement in capacity, speed of learning, interference, and the number of simultaneously active patterns. Clearly, these changes would improve performance importantly. And for each of the aspects cognitive developmental studies can be found that show evidence for these improvement during development (see for example Siegler, 1996; van der Molen and Ridderinkhof, 1998). Moreover, this improvement could result in more powerful behaviour, which is our second criterion. However, these predictions as such are very general and only become meaningful in a detailed simulation study of ART applied to a particular developmental phenomenon.

According to the third criterion, Exact ART should learn rules and conceptual representations. Most neural networks, such as feed-forward PDP-networks, are not naturally suited for modelling higher cognitive tasks (Raijmakers *et al.*, 1996; Quinlan *et al.*, in press). ART networks have some promising applications of learning and abstracting higher-order rules (Anumolu *et al.*, 1997; Levine, 1995; Tan, 1995) but still have to prove themselves in psychological applications.

In sum, in the present chapter, we have tried to explain how the theoretical starting point of modelling a developmental transition by a bifurcation generates a series of constraints or criteria for neural network models of cognitive developmental transitions. Neurological models of network development are consistent with these constraints and provide additional constraints. The ultimate challenge is to define a biologically plausible neural network model of a cognitive developmental transition that escapes Fodor's learning paradox.

Acknowledgements

The research of Dr M. E. J. Raijmakers has been made possible by a fellowship of the Royal Netherlands Academy of Art and Sciences.

Glossary

Bifurcation Roughly speaking, a bifurcation is a qualitative change in an attractor's structure as a parameter (i.e., the control parameter) is smoothly varied. The attractor's structure consists of the equilibrium states. For example, a fold bifurcation changes the number of stable and instable equilibrium states.

Content addressable memory In a content addressable memory, the content of the input is used to retrieve the stored information. Hence, information is only retrieved by the presence of the input. In ART, the stored information is category information. Categories are represented by stable equilibrium states of the SNN (shunting neural network; see text). Retrieval of category information is a parallel process, which takes place by competition between neural populations. Category information is stored in adaptive weights.

Dynamical system A deterministic dynamical system is a system of which the evolution in time is completely defined by a set of equations and its initial state. Differential equations can define dynamical systems: $\frac{dx}{dt} = F(\Gamma, X)$ where dx/dt is the change of x over time. X is a vector of elements x representing the behaviour of the system, Γ is a vector of control parameters, and F is a function of Γ and X. In case F is non-linear in X, the system is a non-linear dynamical system.

Equilibrium state An equilibrium state is a state of the system where the change over time is zero. dynamical system: $\frac{dx}{dt} = 0$, that is $F(\Gamma, X) = 0$ When a dynamical system is perturbed, it moves to a stable equilibrium state and away from an unstable equilibrium state.

Hysteresis Hysteresis represents the history dependence of dynamical systems. By variation of a control parameter, a dynamical system can switch from one equilibrium state to another equilibrium state. With Hysteresis, the direction of change of the control parameter determines at which value of the control parameter the system switches. More precisely, hysteresis means a lagging in the switch due to a changing control parameter.

Notes

1. The match, R, is calculated as follows: $R = 1 - \frac{1}{2}\sum_{i=1}^{N}|u_i - x_i|$. u_i is element i of the activity vector in F0 and x_i is element i of the activity vector in F1. Both vectors have length N.
2. ART3 and Distributed ARTMAP can also form distributed code (Carpenter & Grossberg, 1991; Carpenter *et al.*, 1998). As is pointed out by Carpenter, if distributed codes are learned, there is a trade-off between adaptability and stability of learned categories, which does not occur in local coding.

References

Anderson, P, Gross, GN, Lomo, T and Sveen, O (1969). Participation of inhibitory and excitatory interneurons in the control of hippocampal cortical output. In M. Brazier, (ed.) *The interneuron*, pp. 415–432. Los Angeles, CA: University of California Press.

Anumolu, V, Bray, NW and Reilly, KD (1997). Neural network models of strategy development in children. *Neural Networks,* 10, 7–24.

Bereiter, C (1985). Toward a solution of the learning paradox. *Review of Educational Research,* 55, 201–226.

Bloom, P and Wynn, K (1994). The real problem with constructivism. *Behavioral and Brain Sciences,* 17, 707–708.

Borisyuk, RM and Kirillov, AB (1992). Bifurcation analysis of a neural network model. *Biological Cybernetics,* 66, 319–325.

Brainerd, CJ (1978). The stage question in cognitive–developmental theory. *Behavioral and Brain Sciences,* 1, 173–213.

Buckingham, D and Shultz, TR (2000). The developmental course of distance, time, and velocity concepts: A generative connectionist model. *Journal of Cognition and Development,* 1, 305–345.

Carpenter, GA and Grossberg, S (1987a). A massively parallel architecture for a self-organizing neural pattern recognition machine. *Computer Vision, Graphics and Image Processing,* 37, 54–115.

Carpenter, GA and Grossberg, S (1987b). ART 2: Self-organization of stable category recognition codes for analog input patterns. *Applied Optics,* 26, 4919–4930.

Carpenter, GA and Grossberg, S (1991). ART 3: Hierarchical search using chemical transmitters in self-organizing pattern recognition architectures. In GA Carpenter and S Grossberg (eds)*Pattern recognition by self organizing neural networks,* pp. 453–499. Cambridge, MA: The MIT Press.

Carpenter, GA, Grossberg, S and Rosen, DB(1991). Fuzzy ART: Fast stable learning and categorization of analog patterns by an adaptive resonance system. *Neural Networks,* 4, 759–771.

Carpenter, GA, Milenova, BL and Noeske, BW(1998). Distributed ARTMAP: A neural network for fast distributed supervised learning. *Neural Networks,* 11, 793–813.

Dromi, E (1987). *Early lexical development.* Cambridge: Cambridge University Press.

Eccles, JC, Ito, M and Szentagothai, J (1967). *The cerebellum as a neural machine.* New York: Springer.

Elman, JL, Bates, EA, Johnson, MH, Karmiloff-Smith, A, Parisi, D and Plunkett, K (1996). *Rethinking innateness: A connectionist perspective on development.* Cambridge, MA: A Bradford book, MIT Press.

Fahlman, SE and Lebiere, C (1990). The cascade-correlation learning architecture. In D Touretzky (ed.) *Advances in neural information processing systems, Vol. 2*, pp. 524–532. San Mateo, CA: Morgan Kaufman.

Fodor, JA (1980). On the impossibility of acquiring more powerful structures. In M Piattelli-Palmarini, J Piaget and N Chomsky (eds) *Language and learning: The debate between Jean Piaget and Noam Chomsky*, pp. 142–162. Cambridge, MA: Harvard University Press.

Fodor, JA (1981). *Representations: philosophical essays on the foundations of cognitive science.* Cambridge, MA: MIT Press.

Földiák, P and Young, MP (1995). Sparse coding in the primate cortex. In MA Arbib (ed.) *The handbook of brain theory and neural networks*, pp. 895–898. Cambridge, MA: MIT Press.

Gilmore, R (1981). *Catastrophe theory for scientists and engineers.* New York: Oxford University Press.

Grossberg, S (1976). Adaptive pattern classification and universal recoding: feedback, expectation, olfaction and illusions. *Biological Cybernetics*, 23, 187–202.

Grossberg, S (1988). Non-linear neural networks: Principles, mechanisms, and architectures. *Neural Networks*, 1, 17–61.

Guckenheimer, J and Holmes, P (1993). *Non-linear oscillations, dynamical systems, and bifurcations of vector fields*, 4th edn. New York: Springer.

Hosenfeld, B, van der Maas, HLJ and van den Boom, DC (1997). Indicators of discontinuous change in the development of analogical reasoning. *Journal of Experimental Child Psychology*, 64, 367–395.

Jansen, BRJ and Van der Maas, HLJ (1997). Statistical test of the rule assessment methodology by latent class analysis. *Developmental Review*, 17, 321–357.

Jansen, BRJ and Van der Maas, HLJ (2001). Evidence for the phase transition from rule I to rule II on the balance scale task. *Developmental Review*, 21, 450–494.

Kandall, ER, Schwartz, JH and Jessell, TM (1991). *Principles of neural science, 3rd edn.* CT: Appleton and Lange.

Kelso, JAS, Ding, M and Schöner, G (1992). Dynamic pattern formation: A primer. In J Mittenthal and A Baskin (eds) *Principles of organization in organisms. SFI Studies in the Sciences of Complexity, Proceedings Vol. XIII.* : Addison-Wesley, Longman: New York.

Kendler, TS (1995). *Levels of cognitive development.* Mahwah, NJ: Lawrence Erlbaum.

Laurence, S and Margolis, E (2002). Radical concept nativism. *Cognition*, 86, 22–55.

Lawson, AE and Staver, JR (1989). Toward a solution of the learning paradox: Emergent properties and neurological principles of constructivism. *Instructional Science*, 18, 169–177.

Levine, DS (1995). Learning and encoding higher-order rules in neural networks. *Behavior Research Methods Instruments and Computers*, 27, 178–182.

Levine, DS (2000). *Introduction to neural and cognitive modeling*,2nd edn. Mahwah, NJ: Erlbaum.

Marcus, GF (1998). Can connectionism save constructivism? *Cognition*, 66, 153–182.

Mareschal, D and Shultz, TR (1996). Generative connectionist networks and constructivist cognitive development. *Cognitive Development,* 11, 571–603.

Mareschal, D and Shultz, TR (1999). Development of children's seriation: A connectionist approach. *Connection Science: Journal of Neural Computing, Artificial Intelligence and Cognitive Research,* 11, 149–186.

McClelland, JL (1989). Parallel distributed processing: Implications for cognition and development. In MGM Morris (ed.) *Parallel distributed processing: Implications for psychology and neurobiology*, pp. 8–45. Oxford: Clarendon Press.

Molenaar, PCM (1986). On the impossibility of acquiring more powerful structures: A neglected alternative. *Human Development,* 29, 245–251.

Molenaar, PCM and Raijmakers, MEJ (2000). A causal interpretation of Piaget's theory of cognitive development: Reflections on the relationship between epigenesis and non-linear dynamics. *New Ideas in Psychology,* 18, 41–55.

Piaget, J and Inhelder, B (1969). *The psychology of the child,* 8th edn. New York: Basic Books.

Preece, PFW (1980). A geometric model of Piagetian conservation. *Psychological Reports,* 46, 143–148.

Purves, D (1994). *Neural activity and the growth of the brain.* Cambridge: Cambridge University Press.

Quartz, SR (1993). Neural networks, nativism, and the plausibility of constructivism. *Cognition,* 48, 223–242.

Quinlan, PT (1998). Structural change and development in real and artificial neural networks. *Neural Networks,* 11, 577–599.

Quinlan PT, van der Maas HLJ, Jansen, BRJ, Booij O and Rendell, M (In preparation). Re-thinking stages of cognitive development: An appraisal of connectionist models of the balance scale task. *Cognition*, in press.

Raijmakers, MEJ (1997). Is the learning paradox resolved? *Behavioral and Brain Sciences,* 20, 573–574.

Raijmakers, MEJ and Molenaar, PCM (1997). Exact ART: A complete implementation of an ART network. *Neural Networks,* 10, 649–669.

Raijmakers, MEJ, van der Maas, HLJ and Molenaar, PCM (1997). Numerical bifurcation analysis of distance-dependent on-centre off-surround shunting neural networks. *Biological Cybernetics*, 75, 495–507.

Raijmakers, MEJ, van Koten, S and Molenaar, PCM (1996). On the validity of simulating stagewise development by means of PDP networks: Application of catastrophe analysis and an experimental test of rule-like network performance. *Cognitive Science,* 20, 101–136.

Rodriquez, P (2001). Simple recurrent networks learn context-free and context-sensitive languages by counting. *Neural Computation,* 13, 2093–2118.

Roth, WM (2000). Artificial neural networks for modeling knowing and learning in science. *Journal of research in science teaching,* 37, 63–80.

Shultz, TR, Mareschal, D and Schmidt, C (1994). Modeling cognitive development on balance scale phenomena. *Machine Learning,* 16, 57–86.

Shultz, TR, Schmidt, WC, Buckingham, D and Mareschal, D (1995). Modeling cognitive development with a generative connectionist algorithm. In TJ Simon and GS Halford (eds) *Developing cognitive competence: New approaches to process modeling*, pp. 205–261. Hillsdale, NJ: Lawrence Erlbaum.

Siegler, RS (1978). The origin of scientific reasoning. In RS Siegler (ed.) *Children's thinking: What develops?*, pp. 109–150. Hillsdale, NJ: Wiley.

Siegler, RS (1981). Developmental sequences within and between concepts. *Monographs of the Society for Research in Child Developmentelopment,* 46.

Siegler, RS (1996). Unidimensional thinking, multidimensional thinking, and characteristic tendencies of thought. In AJ Sameroff and MM Haith (eds) *The five to seven year shift: The age of reason and responsibility*, pp. 63–84. Chicago, IL: University of Chicago Press.

Sirois, S and Shultz, TR (1998). Neural network modeling of developmental effects in discrimination shifts. *Journal of Experimental Child Psychology*, 71, 235–274.

Stefanis, C (1969). Interneuronal mechanisms in the cortex. In M. Brazier (ed.) *The Interneuron*, pp. 497–526. Los Angeles, CA: University of California Press.

Thom, R (1975). *Structural stability and morphogenesis*. Reading, MA: Benjamin.

Thompson, LA (1994). Dimensional strategies dominate perceptual classifications. *Child development,* 65, 1627–1645.

Tan, AH (1995). Supervised adaptive resonance theory and rules. In LC Jain, B Lazzerini and U Halici (eds) *Innovations in ART neural networks*, pp. 55–86. Heidelberg: Physica-verlag.

Uylings, HBM, van Eden, CG, Parnavelas JG and Kalsbeek, A (1990). The prenatal and postnatal development of rat cerebral cortex. In B Kolb and RC Trees (eds) *The cerebral cortex of the rat*, pp. 35–76. Cambridge, MA: MIT Press.

Van der Maas, HLJ and Molenaar, PC (1992). Stagewise cognitive development: An application of catastrophe theory. *Psychological Review,* 99, 395–417.

Van der Maas, HL, Verschure, PF and Molenaar, PC (1990). A note on chaotic behavior in simple neural networks. *Neural Networks,* 3, 119–122.

Van der Maas, HL (1998). The dynamical and statistical properties of cognitive strategies: Relations between strategies, attractors, and latent classes. In KM Newell, PCM Molenaar (eds) *Applications of non-linear dynamics to developmental process modeling*, pp. 161–176. Mahwah, NJ: Lawrence Erlbaum.

Van der Molen, MW and Ridderinkhof, KR (1998). Chronopsychophysiology of developmental changes in selective attention and processing speed: A selective review and re-analysis. *Journal of Psychophysiology*, 12, 223–235.

Van Ooyen, A, Van Pelt, J and Corner, MA (1995). Implications of activity dependent neurite outgrowth for neuronal morphology and network development.*Journal of Theoretical Biology,* 172, 63–82.

Van Ooyen, A, Van Pelt, J, Corner, MA and Kater, SB (2003). Activity dependent neurite outgrowth: Implications for network development and neuronal morphology. In A Van Ooyen (ed.) *Modeling neural development*, pp. 111–132. Cambridge, MA: MIT Press.

Van Pelt, J, Wolters, PS, Corner, MA, Rutten, WLC and Ramakers, GJA (2004). Long-term characterization of firing dynamics of spontaneous bursts in cultured neural networks. *IEEE-TBME*, 51(11), 2051–2062.

Van Rijn, H, van Someren, M and van der Maas, HLJ (2003). Modeling developmental transitions on the balance scale task. *Cognitive Science*, 27, 227–257.

Wimmers, RH, Savelsbergh, GJP, Beek, PJ and Hopkins, B (1998). Evidence for a phase transition in the early development of prehension. *Developmental Psychobiology*, 32, 235–248.

Young, MP and Yamane, S (1992). Sparse population coding of faces in the inferotemporal cortex. *Science*, 256, 1327–1331.

Chapter 6

Connectionism in an artificial life perspective: simulating motor, cognitive, and language development

Matthew Schlesinger and Domenico Parisi

Introduction

'Classical' connectionist models of development (e.g., Elman, Bates, Johnson, Karmiloff-Smith, Parisi, and Plunkett, 1996) tend to view behavioural change in the individual as due only to the experience of the individual; i.e., as the outcome of learning by the individual. In contrast, the *Connectionism in an Artificial Life Perspective* (CALP) considers behavioural change as the outcome of adaptation at the level of the population as well as the individual. In this view, individual organisms possess inherited genotypes that specify some aspects of their neural network and behaviour, and in turn, determine how the neural network and behaviour change during life. Inherited genotypes are the present result of a process of change at the population level that has taken place in the preceding generations.

Thus, the CALP studies how populations of individually different networks change in successive generations because of the selective reproduction of individuals and the constant addition of new variant individuals due to sexual recombination and random mutations in inherited genotypes. Given differences in the inherited genotype, development in the individual is not just due to the experiences encountered by the individual. Instead, it results from the interaction between individual experiences and the inherited genotype. From this theoretical perspective, we refer to changes that are mainly (but never exclusively) due to the inherited genotype as 'maturation', while changes that are mainly (but never exclusively) due to experience are called 'learning', and changes in which the role of genetic information and experience seem to be equally important are called 'development'.

Both versions of connectionism can be said to instantiate a (Piagetian) 'genetic epistemology' according to which, in order to understand X, one should try to reconstruct how X has become what it now is. However, unlike 'classical' connectionism, CALP goes beyond Piaget's emphasis on change in the individual because its 'genetic epistemology' perspective includes both phylogeny and ontogeny (Parisi and Schlesinger, 2002).

CALP uses genetic algorithms (Holland, 1975) to study evolution. However, genetic algorithms can be viewed as abstract models of change in general, and as such they can be used in two different ways: (1) they can be used together with learning algorithms to study how evolution *across* individuals creates innate bases for learning and development in the individual, but (2) they can also be used to study development *within* an individual, with the behaviour of successive generations of organisms representing the successive developmental phases within the individual.

Another important difference between 'classical' connectionism and CALP is that individuals are always embodied and situated. 'Classical' neural networks do not have a body and effectively live in a void. More precisely, their only 'environment' is the researcher who decides which input arrives to the network's input units and who evaluates the network's output and provides a teaching input for learning. However, 'natural' learning consists to a large extent of learning how the environment responds to the organism's own actions. This cannot be studied if there is no environment. In contrast, in CALP, a neural network is just one physical component of the organism's entire body. The organism lives and interacts within a physical environment. In a CALP simulation, one not only simulates a neural network and what takes place inside the network, but one also simulates the organism's body, with its specific physical properties such as size and shape and nature and location of sensors and effectors, and the physical environment in which the organism happens to live. The researcher does not decide upon the inputs to the organism's neural networks arbitrarily. They depend on the physical characteristics of the environment and of the organism's body and on the current location of the organism in the environment. This has the important consequence that the network's outputs, in particular the movements controlled by the network, tend to change either the physical relation of the organism's sensors to the environment or the environment itself. Thus, these organisms (at least in part) determine their own inputs (Nolfi and Parisi, 1993).

Because they are embedded within an environment, CALP neural networks may be called 'ecological' networks. Supervised learning algorithms such as the back-propagation algorithm are rightly criticized as ecologically implausible because in many cases it is not clear where the teaching input that specifies the correct output comes from (Parisi, 1996; Parisi, Cecconi, and Nolfi, 1990). This criticism appears to be justified but not in all cases. For example, if a neural network is learning to predict the next state of a changing environment or the consequences of the organism's own actions on the environment, the next input arriving from the environment can be used as teaching input in an ecologically plausible way. Another example is a neural network that learns to imitate the output of another neural network in response to some input. The 'imitator' uses the 'model's' output as teaching input in a perfectly ecological manner.

In this chapter we provide an overview of modelling within the CALP framework. We then describe three simulations that illustrate the approach with examples from motor, cognitive, and language development.

Modelling overview

In this section we provide a general introduction for non-specialists into the technical aspects of designing and studying a CALP simulation. For additional details, the interested reader may also consider Schlesinger and Parisi (2001b) and Schlesinger (2004); similarly, a comprehensive introduction with several examples is provided by Parisi, Cecconi, and Nolfi (1990).

We begin the overview by first describing the key ingredients of a simulation, while stressing the integrative aspects of the model. Next, we focus on the role of the autonomous agent or organism, and different levels at which the organism can be designed and studied. We then consider the task of designing an environment for our simulated organisms to inhabit. A related problem concerns the choice of how organisms will adapt to their environment—evolve, develop, etc.—that is, the learning algorithm.

Basic principles

The Artificial Life framework that we employ is general and versatile enough to allow for formulating and investigating a wide range of research questions, varying across both a spatial and temporal scale. A particularly useful aspect of our approach is that it is consistently multilevelled. For example, a typical simulation will include at least four distinct spatiotemporal scales:

1. a genetic level, including biological structures such as a genome,
2. a neuroanatomical or neurophysiological level, represented by an artificial neural network,

3. an ontogenetic level, at which an individual organism spends its life, and finally
4. a phylogenetic level, across which a population of organisms adapts on an evolutionary timescale.

Note that it is ultimately up to the model-builder to decide which of these levels should be included in their simulation, and if so, how much detail (i.e., biological and physical realism) needs to be incorporated into the model. To illustrate the flexibility of this approach, consider just a few examples of the kinds of simulations that we have investigated, and which vary widely in scope. For instance, one line of research (Schlesinger and Parisi, 2001b; Schlesinger, 2003; Schlesinger and Young, 2003) simulates infants' perceptual activity (e.g., eye movements, visual scanning, etc.) in the specific context of an object permanence study (Baillargeon, 1986). This work has a relatively narrow focus on microgenetic learning experiences that occur in individual children as they participate in a particular experiment. Another series of studies investigates the development of reaching (Schlesinger and Parisi, 2001a; Schlesinger, Parisi, and Langer, 2000). Although operating on a longer developmental timescale, this work also focuses on ontogenetic changes within the individual. Finally, we have also studied numerous simulations which investigate changes in a population of artificial organisms, including maturation, learning, development, and evolution (e.g., Parisi, Cecconi, and Nolfi, 1990; Nolfi and Parisi, 2002).

Constructing the organism

The concept of an autonomous, embodied organism plays a central role in our research. However, it is important to note that our use of the term embodiment extends well beyond the idea of possessing a physical body—a torso, arms, legs, etc. Instead, consistent with the notion of a multilevelled model, we conceptualize our organisms as embodied on several, simultaneous levels. Each level of embodiment provides a distinct set of physical constraints that influence developmental and evolutionary processes in a unique manner.

One level of embodiment is the physical shape and size of the organism, including how they occupy space, and where possible, how they move through it. At the same spatiotemporal scale, another closely related level includes both the sensors (visual, auditory, haptic, etc.) that provide information about the current state experienced by the organism, as well as effectors or motor systems (arms, legs, eyes, mouth, etc.) for manipulating or changing the environment. At a more fine-grained level, there is also an artificial nervous system, which not only links the sensors to the effectors (i.e., input to output), but also provides a physical structure upon which learning, development, and evolution can operate (e.g., by changes in network connection weights, network topology, etc.). In virtually all of our simulations, we have employed an artificial neural network as a model of the nervous system.

Finally, in models that simulate evolutionary change, the model-builder must also construct a genome. Note that virtually any characteristic of the organism can be encoded in the genome, including not only the 'congenital' properties of the neural network (i.e., those acquired at birth, for example the network architecture, the initial connection weights, etc.), but other inherited physical characteristics that may either not emerge at birth, or may change over time (size, shape, growth rates, learning rates, etc.). While there are a variety of techniques for representing these characteristics in a genetic code, a common strategy is to use binary strings of varying lengths to encode each property. For example, the strength of a connection between two units in the neural network can be encoded as a string of eight ones and zeros. The entire network is then represented genetically by this collection of successive binary strings, analogous to a string of DNA.

Designing the environment

In most of our work, we study simulated environments, though some evolutionary robotics studies include real environments (e.g., Nolfi, Floreano, Miglino, Mondada, 1994). Note that the process of designing the physical environment in which the organism is situated, in many ways, compliments how the organism is constructed. For example, if the organism has visual sensors, then there should naturally be features in the environment that can be seen and possibly discriminated, such as objects of different shapes, sizes, and colours. Similarly, if the organism is capable of locomotion, then the environment needs to be large enough, in relation to the organism, so that movement is possible. This synergy between the organism and the environment is no accident, of course, but instead follows logically from the idea of an embodied and situated organism.

In much of our work, we employ a minimalist principle in designing the environments that our organisms inhabit. That is, we try to avoid the mistake of making the environment 'too real', for several reasons. First, overly complicated environments are not only difficult to program, but also consume large amounts of computing time and resources. A second and related concern is that unnecessarily complex environments often cloud or confuse the key relations we hope to study in simulation. Indeed, one of the main points of simulating a given phenomenon is to investigate it while stripping away extraneous details.

On the other hand, a critical feature of the environment is that it should incorporate as many physical and/or geometric constraints as necessary. In the case of our model of reaching, for example, this means carefully working out the equations which describe the position of the organism, and how a new position or posture is achieved by a combination of arm and eye movements. This kinematic model can be made more complicated by the addition of dynamic forces (gravity, inertia, etc.), which interact with each other in complicated ways. Similarly, models that employ vision as one of the organism's sensory modalities may have to account for various geometric properties of light propagation, such as occlusion and reflection.

Selecting the learning algorithm

Broadly construed, the learning algorithm is a set of computational rules which evaluate the behaviour of the organism (or population of organisms), and produce a set of changes in the organism's neural network such that desired behaviours improve or become more likely. From an Artificial Life perspective, the learning algorithm takes on additional meaning. Not only is it a technique for studying adaptive behaviour, but different algorithms can also be interpreted as adaptation over different timescales. For example, a genetic algorithm may operate by selecting the 'most fit' organisms in a generation—those most successful on a given task, such as finding food—and using them to produce the next generation of organisms. Copying errors (i.e., mutations) in the neural networks of the offspring guarantee continued variability in performance, from which more successful organisms may be selected. This scenario, of course, is intended to be an implementation of neo-Darwinian processes via simulation.

It is important to remember that using a genetic algorithm is rarely intended to be a literal representation of how evolution occurs in biological organisms, nor more specifically, of how neural structures arise and change over time. Instead, as we have argued elsewhere (e.g., Schlesinger, 2004), genetic algorithms play a more metaphorical role within CALP, providing a computational framework that (a) operates at the evolutionary timescale, (b) creates the opportunity to simulate innate—i.e. genetically specified—behaviours, and (c) is versatile enough to allow multiple levels of change to be systematically studied and compared (e.g., related changes at the behavioural, neural, and genetic levels; see Miglino, Nolfi, and Parisi, 1996).

We have also investigated other learning algorithms on different timescales. As a proxy for learning during the individual lifespan, for example, we have implemented the back-propagation-of-error algorithm in a population of artificial organisms. An interesting question arises when multiple algorithms are combined: For example, what happens when we simulate a population that evolves via a genetic algorithm, but is also comprised of individuals who learn via back-prop? Our work suggests that adding the capacity for learning or development, over and above the changes produced by evolution, can accelerate or facilitate the appearance of the evolved behaviour (Parisi, Cecconi, and Nolfi, 1990; Hinton and Nolan, 1987). Interestingly, in some cases learning may facilitate evolution even when the learned behaviour (e.g., learning to predict the sensory consequences of planned movements) is different from the evolved behaviour (e.g., foraging for food).

Our view, more generally, is that there is a wide array of learning methods appropriate for connectionist models, including not only genetic algorithms and backprop, but also reinforcement learning, autoassociators, Hebbian learning, and adaptive resonance theory (for recent reviews, see Parisi and Schlesinger, 2002; Schlesinger and Parisi, 2004). To our knowledge, only the genetic algorithm approach seems well-suited for simulating evolutionary processes, while each of the others might be used interchangeably for investigating ontogenetic processes such as learning, maturation and development.

In order to illustrate how the concept of an embodied and situated organism can play a central role in developmental research, we next present a brief survey of our work from three developmental domains: motor development, cognitive development and language development.

Motor development

Unlike models of cognitive processes (e.g., object recognition or categorization) which do not necessarily need to simulate an embodied organism, models of motor activity are almost inevitably embodied. This is not only because there is a broad consensus among researchers that motor development is an interactive process (e.g., Berthier, 1996; von Hofsten, 1984; Thelen, Corbetta, Kamm, Spencer, Schneider, and Zernicke, 1993), but also because motor activity requires both sensing one's current state in the environment (the location of the hands with respect to a nearby object) and also acting on the environment (reaching, moving, looking, vocalizing, etc.).

As an example of how embodiment plays a central role in motor development, consider the problem of learning to reach. Within approximately their first three months, young infants acquire the ability to move their hands efficiently toward desired objects (efficient grasping, on the other hand, develops more slowly). This ability involves the coordination of at least two major sensory systems—vision and proprioception—and three major motor systems—movement of the head, eyes, and arms. What physical constraints play a role in the acquisition of this new skill?

We addressed this question by designing and studying a population of artificial neural networks, which represent the neural control systems that underlie the development of hand-eye coordination in young infants (Schlesinger, Parisi, and Langer, 2000). The model includes constraints on the development process that are both embodied and situated. First, the simulated physical body includes a monocular visual system, a two-segment arm (with a ball joint at the shoulder and a hinge joint at the elbow), and a feedforward neural network (see Box 6.1). One geometric constraint imposed by this design is that rotation of the body axis or shoulder joint causes the arm end point to move, even when the elbow joint is held in place. As we see below, a developmental consequence is that control of the arm proceeds in a centre-to-periphery, or proximodistal pattern, with consistent use of the shoulder joint emerging before use of the elbow joint. Second, a related, situated constraint follows from placing the organism in an environment in which it can rotate its visual field 360°. Consequently, when neither the hand nor the target object are visible, the organism can learn to either (a) first find the target and then bring the hand into the visual field (i.e., the more efficient strategy), or (b) first find the hand, and then 'drag' it around while searching for the target. Indeed, the model reliably acquires the former strategy, mirroring the developmental pattern in human infants (e.g., Clifton, Rochat, Robin, and Berthier, 1994).

Several important conclusions emerge from this work. First, note that the physical design of our organism includes a two-segment arm (i.e., an upper-arm and forearm) with rotation limits comparable to a human arm; the organism can also rotate its torso while reaching. Thus, the organism effectively has three movement degrees of freedom (DOF), while the target object is located on a two-dimensional plane, leaving the organism with one redundant DOF (this is known as the 'DOF problem', see Bernstein, 1967). Instead of using all three DOF, the model instead quickly learns to 'freeze' the shoulder joint and use only torso and elbow rotations to reach a target. Interestingly, the strategy of freezing redundant joints is a common phenomenon in adult novices as they acquire new motor skills, for example learning to ski (see Vereijken, Emmerik, Whiting, and Newell, 1992).

A second major result that emerges from our reaching model concerns how opposing pairs of muscles (i.e., flexors and extensors) that are used to rotate each joint are coordinated. In the model, we observe a proximodistal pattern of development, in which the muscles controlling the shoulder joint become coordinated before those controlling the elbow joint. This result not only corroborates kinematic analyses of infants' reaching movements (i.e., changes in hand position and velocity, e.g., von Hofsten, 1984; Trevarthen, Murray, and Hubley, 1981), but also suggests that the initial stiffness or co-contractions

Box 6.1 The development of hand-eye coordination

Our econet (ecological neural network) model of reaching (Schlesinger, Parisi, and Langer, 2000) was designed to simulate the development of hand-eye coordination in young infants. A key focus of the model is on the role of trial and error learning mechanisms during the development of reaching. Accordingly, we implemented a population-based approach in which a number of alternative movement patterns are produced and evaluated, with selection and reproduction of the most successful movements.

This work not only highlights the role of trial and error learning, but it is also loosely based on the concept of a population of motor neurons, in which subsets of neurons each code a particular movement direction (Georgopolous, Schwartz, and Kettner, 1986). Consequently our model represents an individual child by simulating a population of neural networks, each of which 'competes' for the control of the individual's eye and arm movements.

Nevertheless, we should stress in this case that not only the population-based approach, but also our use of a genetic algorithm are intended as a metaphor or proxy for the trial and error strategies employed by an individual child, rather than as an explicit assumption regarding the actual neural mechanisms involved (for a related discussion, see Schlesinger, 2004).

As Figure 6.1a illustrates, each econet is initially represented by a binary string, which specifies the physical characteristics of the econet. In the first generation of econets, each string is a random sequence of 1's and 0's. The string is then used to 'build' the econet, by converting portions of the string into corresponding characteristics. For example, initial connection weights in the econet's neural network (not shown here) are encoded as eight-bit segments in the binary string.

Note that other physical features, such as size and shape, might also be encoded in the same way. By repeating this process a number of times, as Figure 6.1b illustrates, an entire population of econets is created. Because each is created with a different binary string, the first generation of econets is composed of unique individuals.

The next step in the modeling process involves evaluating the fitness of each individual. This is accomplished by placing each econet in its environment—in this case, a simple world with one object (see Figure 6.1c)—and giving it several trials to reach for the target object. At the

Box 6.1 *(continued)*

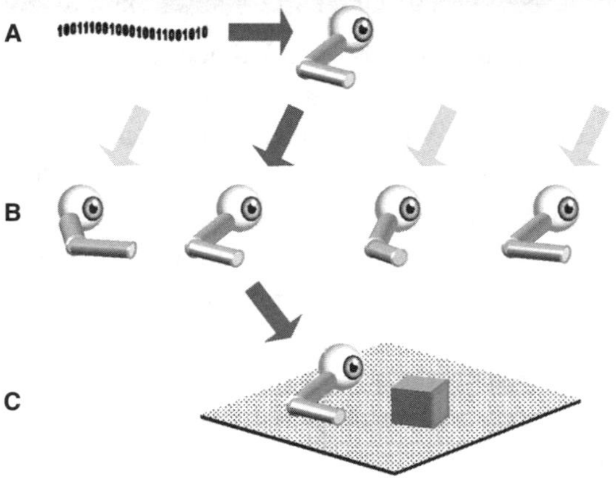

Fig. 6.1 The reaching model. (A) A binary string is used to generate the physical characteristics (i.e., phenotype) of each organism. (B) The first generation is composed of a population of organisms, each created by a random binary string. (C) Each organism is evaluated by placing it in the environment, and testing its ability to make contact with a target object. (See Box 6.1 for details.)

start of each trial, both the econet and the object are randomly positioned. The econet's movements are determined by its neural network, which receives visual, proprioceptive, and tactile sensory inputs and generates eye and arm movements. Trials end when the econet either reaches the target, or fails to reach it after a predetermined number of movements. After each successful trial the econet's fitness is increased by 1.

Learning is accomplished in the model by selecting the most successful econets in each generation (i.e., those with the highest fitness), and making multiple copies of each to produce the next generation. Occasional copying errors (i.e., mutations) in the offsprings' binary strings introduce variations in behaviour, with the possibility that some offspring will be more successful than their parents at reaching. By repeating these steps in an iterative fashion, we eventually obtain a population of econets that are successful at reaching efficiently.

observed throughout young infants' arms and shoulders as they begin to reach (Thelen *et al.*, 1993) should diminish first in proximal segments of the infant's body before more distal segments (e.g., in the upper arm before the forearm).

More recently, we have used the reaching model to investigate the sensory encoding strategies that are available to infants for locating the position of their arms in space, and for moving their hand toward a desired target (Schlesinger and Parisi, 2001a). A number of studies suggest that, contrary to earlier theories, infants do not learn to reach by visually guiding their hand toward a target (e.g., Clifton, Muir, Ashmead, and Clarkson, 1993; Clifton, Rochat, Robin, and Berthier, 1994). Instead, by the onset of reaching, infants are equally successful at reaching for a target, whether or not their hand is visible (i.e., under normal illumination vs. an illuminated target in a darkened room). This and other empirical observations (e.g., infants do not attend to their hands as they reach) suggest that vision and proprioception become coordinated through a modality other than vision.

In an extension of our original reaching model (Schlesinger and Parisi, 2001a), we hypothesized that tactile input, made during contact with the target, might provide a third sensory channel that helps 'triangulate' vision of the target with proprioception from the arm. A series of simulations provided strong support for our hypothesis. First, we compared learning across two conditions (i.e., in two populations of artificial neural networks): in one condition, a tactile sensor becomes active during contact with a target, while in the second condition the tactile sensor is permanently deactivated. Performance was superior, across a number of measures, when tactile input was available. Specifically, in this condition (1) learning was more rapid and reached a higher level of performance, (2) latency to contact with the target was shorter, and (3) after reaching the target, more sustained contact with the target was achieved.

Second, we designed a training scenario analogous to the Clifton *et al.* (1993) study, in which reaching under normal lighting conditions was compared to reaching toward luminous targets in the dark. In particular, we simulated a condition, comparable to the first described above (i.e., with tactile input from the hand), but with one important change: the hand and arm were made invisible, so that while the target could be seen, the hand could only be localized through proprioceptive feedback from the arm. Note that this simulation provides a critical test of our hypothesis. Specifically, if sight of the hand is necessary in the model for learning to reach, then performance should be severely disrupted by making the hand invisible. Alternatively, if contact with the target during reaching is sufficient for coordinating vision and proprioception, then making the hand invisible should not affect performance.

Our results provided unequivocal support for our hypothesis. That is, we found that even when the hand is made invisible, there are no disruptions in either the learning trajectory or final performance. Effectively, the model learns equally well to coordinate vision and proprioception, regardless of whether the hand is visible. A number of empirical predictions follow from these results. First, if touching an object acts like a 'reward function' that links vision and proprioception, we should predict that postures (i.e., head, eye, and arm configurations) that are near to those during successful reaching trajectories should be more salient to infants, and therefore easier for them to discriminate than postures that are far from reaching trajectories. Second, if infants' visual–proprioceptive coordination is perturbed for example by wearing displacing prisms that make objects appear in non-veridical locations (see McDonnell and Abraham, 1979), we should also predict that infants will adapt more rapidly as they reach for real objects, that is, when there is tactile feedback, than when they reach for virtual objects (e.g., holographic images).

Cognitive development

Most computational models of cognitive development do not employ an embodied or situated approach (but see Jones, Ritter, and Wood, 2000; Sporns, Chapter 8 this volume; Thelen, Schöner, Scheier, and Smith, 2001 for recent exceptions). Instead, the prevailing approach is to simulate a specific neural circuit or processing loop, while side-stepping the issue of designing both the physical structure of the organism, as well as the structure of the environment. In this case, the environment is treated as a stream of sensory information, which is pre-parsed or segmented into a set of input patterns. These patterns then comprise a training set, from which a pattern is selected at random and presented to the model. We referred above to this training regime as 'classical connectionism' (Parisi, 1997).

Unfortunately, this approach is somewhat at odds not only with developmental theories that stress the importance of exploration and interaction with the environment (e.g., Gibson, 2000; Piaget, 1952), but also with developmental research that highlights the role of self-directed activity (e.g., Bertenthal, Campos, and Kermoian, 1994; Held and Hein, 1963). A related concern is that many models also focus on the role of internal representations, while often overlooking the contribution of implicit knowledge that is encoded or represented through skillful interaction with the environment (e.g., procedural knowledge, sensorimotor representations, etc.; see Ballard, Hayhoe, Pook, and Rao, 1997; Brooks, 1991; O'Regan, 1992).

In contrast, the notion of an embodied and situated organism provides an ideal framework not only for investigating the role of self-produced activity in cognitive development, but also for identifying a wide array of constraints that influence representations as they emerge early in development. To illustrate our point, consider the development of object permanence, which involves the understanding that an object continues to exist while it is out of sight (Piaget,

1952). As a specific example, imagine an infant confronted with the problem of visually tracking a moving object, which occasionally disappears and then reappears as it passes behind a screen.

One strategy for addressing this problem is to assume that infants form internal representations of hidden objects, and that they use these representations to guide their behaviour (e.g., where to search for the hidden object), when perceptual contact with the object is lost. This approach has inspired a number of connectionist models (e.g., Mareschal, Plunkett, and Harris, 1999; Munakata, McClelland, Johnson, and Siegler, 1997), which seek to explain the development of infants' search behaviours in terms of progressively more detailed and accurate internal representations or mental models.

As an alternative approach, we might assume that infants' earliest representations are not internal 'copies' of external experience, but instead are pragmatic systems of knowledge about how one can act in the current situation, and how ongoing experience may be transformed by those actions (for related accounts, see O'Regan and Noe, 2001; Thelen and Smith, 1994). From this perspective, the developmental task is not to learn to construct a (disembodied, general-purpose) mental image, model, or representation of hidden objects and events, but rather, to construct a set of task-specific representations that link perception with action. Naturally, these links form the basis not only for adaptive, prospective, goal-directed behaviour, but also for knowledge about the physical world (including the concept of object permanence). Note that this knowledge therefore is always embedded in action, and is situated and embodied.

In order to illustrate how this alternative approach might work in practice, we designed a connectionist model of an infant as it learns to visually track moving objects (Schlesinger and Barto, 1999; Schlesinger and Parisi, 2001b). In contrast to our other models, the tracking model focuses on change at the individual or ontogenetic level, during a relatively brief time span (e.g., during an experimental session). There are three important aspects of the tracking model.

First, it implements a relatively simple, but nevertheless embodied design by simulating an oculomotor control system, including: (1) an 'eye' with a two-dimensional retina, (2) a feedforward neural network, and (3) a motor system that drives eye movements, allowing the model to shift its gaze from one location to another (i.e., to control its visual input over time). Second, there are neither built-in memory systems (e.g., a delayed input line, recurrent connections from one layer in the network to another, etc.) nor sensory prediction systems. In other words, the model's ability to act prospectively (e.g., to anticipate when and where a hidden object will reappear) is not based either on input that has been temporally coded, or on an internal model that computes past or future states. Third, and perhaps most importantly, the overall design of the model is based on bottom-up principles; in particular, there is no built-in knowledge or abilities, and therefore, no top-down influence on the model's perceptual processing from previous experiences or expectations.

We began our evaluation of the tracking model by assessing its ability to keep a moving target within view. Specifically, we trained the model to track the target across a variety of conditions (e.g., normal tracking, tracking with a brief occlusion of the target, etc.). The performance of the model during training, as well as its generalization to novel tracking events, replicated several key aspects of visual tracking development in young infants (e.g., Aslin, 1987; Bremner, 1985). First, the model learns to track fully visible targets before tracking targets that are briefly occluded. Second, as it learns to track occluded targets, the model's initial strategy is to continue to 'fixate' the location where the target was last seen (i.e., at the point where it becomes occluded). Third, like human infants, as the model tracks occluded targets it also learns to generate anticipatory eye-movements or saccades to the location where the occluded target reappears. Finally, the model also generates a pattern of gaze-shift errors comparable to those produced by nine-month-old infants (Berthier *et al.*, 2001), when the target encounters an obstacle during occlusion (Schlesinger and Parisi, 2001b).

A key result of the tracking model is that when the target is occluded, like human infants, the model generates a series of anticipatory eye movements, toward the location where the target will reappear (Johnson, Amso, and Slemmer, 2003). However, given the design of the model, it is clear that this prospective or future-oriented behaviour is not the result of an explicit prediction or computation, but is instead simply an adaptive by-product of acquiring an effective tracking strategy.

Given these initial positive findings, we wondered how the tracking model might inform the debate on infants' early object knowledge (e.g., Baillargeon, 1999; Haith, 1998; Smith, 1999; Spelke, 1998). One of the questions raised by this debate is what kind of knowledge is tapped by methods such as the violation-of-expectation (VOE) paradigm, which measure infants' looking time during possible versus magical or physically impossible events? In particular, why do infants often look longer at impossible events (e.g., Spelke, 1985)? We addressed this question by elaborating the tracking model in a number of ways, so that it could 'watch' a series of possible and impossible events comparable to those shown to infants in a study that investigates the development of object permanence (Baillargeon, 1986).

In Baillargeon's 'car study', infants see a car that rolls down a track, behind a screen, and then out the other side (for additional details, see Box 6.2). After several repetitions of this event, infants then watch a series of possible and impossible events, in which an obstacle is placed either behind or on the track, and is then hidden by the screen. After the obstacle is hidden, the car then appears and moves along the track as before. Baillargeon found that infants looked significantly longer when the obstacle was placed on the track (i.e., during the impossible event), suggesting that they were surprised to see the car reappear when its path was obstructed by a hidden object.

Box 6.2 The development of object permanence

Baillargeon's 'car study' (Baillargeon, 1986; Baillargeon and DeVos, 1991) was designed to show that infants are able to mentally represent hidden objects and their properties. Figure 6.2a illustrates the general sequence of events in this study. Infants begin by watching the habituation event, in which (a) a screen is lifted and then lowered, and then (b) a car appears and rolls down a track, passing behind the screen. After several repetitions of the habituation event, infants watch two test events. During both the possible and impossible test events, a box is now revealed when the screen is lifted; the box is behind the track during the possible event, but on the track during the impossible event. During both events the car rolls along the track as it did during the habituation event.

Baillargeon found that young infants looked significantly longer at the impossible event, and interpreted this result as evidence that infants were surprised to see the car reappear when its path was obstructed by the box. She concluded that infants not only knew that the box and car existed while occluded, but also knew that both objects could not be in the same place at the same time.

Our work raises the possibility of another interpretation. Specifically, infants might look longer at the impossible event because the location of the box (i.e., in the car's trajectory) draws the infant's attention, causing them to scan the event in a novel way, and therefore making the impossible event more salient or interesting.

We investigated this alternative account by modeling infants' gaze patterns during the car study. First, we created computer-animation events that corresponded to those in the original car study. Figure 6.2b illustrates selected frames from these animation sequences. Second, we designed a multi-layer network, which 'watches' the habituation event and learns to track the movements of the car.

Box 6.2 *(continued)*

As Figure 6.2c illustrates, each animation frame is processed into two input streams. In the 'peripheral' stream, a low-resolution pattern is obtained by averaging large areas of pixels in the animation frame. In the 'fovea' stream, a high-resolution pattern is obtained by sampling a small region of the input while preserving the pixel values from the animation frame. In addition, a third input identifies the center of the fovea stream (i.e., the fixation point, indicated by a green x in the animation frame) in Cartesian coordinates. Input values are then propagated forward through the network, generating an eye-movement (i.e., a shift in the fixation point); the two banks of output units correspond to horizontal and vertical shifts, respectively, which are superimposed.

During training, the model is rewarded for keeping the car within the fovea. Specifically, a reinforcement-learning algorithm is used to modify the connection weights in the network, increasing the likelihood that the model will track the car (for additional details, see Schlesinger, 2003). After training, the model is then presented with the possible and impossible test events (see Figure 6.2b).

Our gaze-direction model generates a number of empirical predictions concerning infants' gaze patterns during the test events. First, the model predicts that overall, infants will gaze most toward the center (i.e., toward the screen), with fewer fixations toward the right and the left (i.e., toward the car's exit and entry points, respectively). This prediction was confirmed in a replication study of six-month-old infants (Schlesinger and Casey, 2003).

Second, the gaze-direction model also predicts that infants will scan the possible and impossible events differently. In particular, when infants increase their looking time during the impossible event, most of these 'additional' fixations (i.e., those over and above the looking time spent in the possible event) should be toward the centre of the display. Again, this prediction was also supported by the evidence (Schlesinger and Casey, 2003).

These results suggest that infants' gaze patterns may not only provide a more detailed picture of their reactions to possible and impossible events, but they may also be driven in part by perceptually salient aspects of the display. Our long-term goal is to continue to develop a modeling approach that includes both perceptual-processing and knowledge-based systems, which will allow us to ultimately tease these two levels of explanation apart.

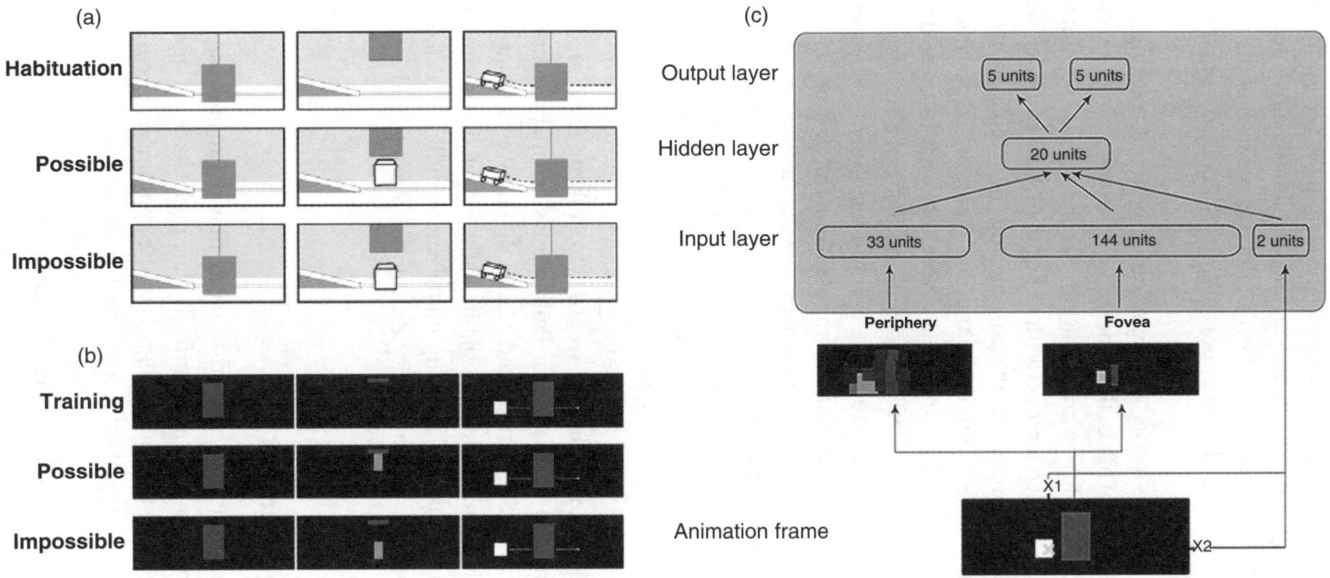

Fig. 6.2 The tracking model. (A) Baillargeon (1986) investigated infants' reactions in the 'car study', in which a car rolls briefly behind a screen; infants watch the habituation event, followed by the possible and impossible test events. (B) The tracking model is presented with analogous, computer-animated events. (C) The architecture of the tracking model's oculomotor control system; each frame of animation is processed along parallel periphery and fovea input streams, leading to the production of a shift in the fixation point. (See Box 6.2 for details.)

We began by first presenting the tracking model with a computer-animation event comparable to the event that infants first watch in the car study. Specifically, the model was trained to fixate the car, and to follow its movement. After training, the model was then presented with animated versions of the possible and impossible events. A number of important results emerged from this work (Schlesinger, 2003).

First, we found that the model was significantly better at tracking the car during the possible event. That is, the appearance of an object in the car's path disrupted the model's ability to track the car. Second, as we monitored the internal activity (i.e., hidden layer activations) of the tracking model, we also found that the model 'represented' the possible and impossible events differently. In particular, relative to the initial training event, the impossible event was significantly more novel or unfamiliar. Note that each of these findings provides an indirect replication of Baillargeon's results, and also suggests why infants may look longer at the impossible event in her study: (1) because the car is more difficult to track during the impossible event, and/or (2) because the impossible event is also more perceptually novel or unfamiliar.

The latest version of the tracking model has generated a variety of empirical predictions regarding infants' moment-to-moment visual activity during VOE studies like the car study (Schlesinger, 2003). For example, the model not only predicts that infants will distribute their gazes differently during the possible and impossible events, but also suggests a specific qualitative pattern of gaze-direction during these events. In order to evaluate these predictions, we recently conducted a replication of Baillargeon's original car study with six-month-old infants (Schlesinger and Casey, 2003). We not only found, like Baillargeon, that infants looked significantly longer at the impossible event, but also found that the tracking model successfully predicted qualitative features of infants' visual activity during the possible and impossible events.

Language development

Artificial Life simulations can be used to study how organisms can acquire an ability to respond to signals appropriately (i.e., understanding or comprehension) and to produce the various signals in the appropriate circumstances (i.e., production). More specifically, Artificial Life simulations can be used to investigate how different classes of signals (words) emerge in the experience of language users and how smaller signals (words) can make up larger signals (sentences).

Imagine an organism whose neural network has visual input units encoding the location and perceptual properties of objects present in the environment and auditory input units encoding the sounds produced by a conspecific. The network has to respond to the inputs by generating the appropriate locomotory movements. The environment contains edible and poisonous mushrooms and each individual mushroom is different from all others (Cangelosi and Parisi, 1998). If an encountered mushroom is sufficiently near to the organism, the organism can perceive both the location and the perceptual properties of the mushroom and therefore it can recognize if the mushroom is edible or poisonous and respond appropriately, i.e., it approaches and eats the edible mushrooms and avoids the poisonous ones. However, because of the organism's sensory limitations, if the mushroom is more distant, only the mushroom's location is perceived by the organism's neural network but not the mushroom's perceptual properties. Hence, in these circumstances the only option open to the organism is to approach every mushroom until the mushroom is sufficiently close that the organism can recognize if it is of the edible or poisonous variety. But of course the resulting behaviour would not be very efficient.

Language can emerge to solve this problem. Another individual, who is nearer to the mushroom or has fewer sensory limitations than the first individual, recognizes the mushroom and generates one of two sounds, X1 and X2, which co-vary with the two categories of mushrooms. The first individual hears the sound produced by the conspecific and responds by approaching the mushroom if it hears the sound co-varying with edible mushrooms and going away from the mushroom if it hears the other sound.

Are the two sounds of this very simple language, X1 and X2, nouns or verbs? There is no real justification for answering one way or another. The two sounds can be paraphrased by using either two English nouns (actually, noun phrases), 'edible mushroom' and 'poisonous mushroom', or two English verbs, 'approach' and 'avoid'. Hence, calling the two sounds either nouns or verbs would be arbitrary.

Now imagine another, somewhat more complex, scenario. The organism has a retina and a two-segment arm, like in the simulations previously described. The environment contains only two objects that have different shapes. At any given time the organism sees both objects at the same time and it hears one of two different sounds, N1 or N2. The organism responds to the input by moving its two-segment arm. When it hears N1, the organism is rewarded if it reaches for the object possessing a given shape, while when it hears N2 it is rewarded if it reaches the object with the alternative shape. Now we would be somewhat justified in saying that the two sounds are nouns, if we define a noun as a sound which co-varies with the particular object on which a given action (reaching, in our case) is executed, independently of the particular action which is executed on the object.

Box 6.3 The early development of language

A primary strength of CALP is its versatility and flexibility. In particular, any given model can easily be modified to study a new developmental question. Take as an example our econet reaching model (see Box 6.1), which we reconfigured to address the problem of acquiring proto-linguistic signaling systems (Parisi, Cangelosi, and Falcetta, 2002).

Imagine a population of artificial organisms with a single arm that lives in an environment which contains different objects that they have to reach or to push/pull in order to survive and reproduce. As Figure 6.3a illustrates, each individual also has an 'ear' which can receive auditory signals. Using a genetic algorithm, the organisms learn in a succession of generations to execute the appropriate action on the appropriate object in response to both the particular object(s) that they see and the particular linguistic signal(s) that they hear. An organism's behaviour is controlled by a neural network (not shown here) with (a) visual input units encoding the perceptual properties of the objects, (b) proprioceptive input units encoding the current position of the organism's arm, and (c) and auditory input units encoding heard linguistic signals, plus output units encoding movements of the arm.

In one simulation (Figure 6.3b) the organisms each time see a single object which can belong to one of two categories and they have to push the objects that belong to one category and to pull the objects that belong to the other category, where the objects are encoded in the input visual units as deviations from two prototype binary patterns. The addition of two linguistic signals that co-vary with the two categories makes the organisms' performance more efficient. However, in this simulation the signals cannot be classified as either nouns or verbs since they co-vary at the same time with both the action to be executed and the category of the object on which the action is to be executed. This is what the first signals understood or produced by children are likely to be.

In a second simulation (Figure 6.3c), the organisms see two objects at the same time, each belonging to one of the two categories and they learn to reach one object if they hear one signal and the other object if they hear the other signal. The two signals can be interpreted as proto-nouns if nouns are defined as signals that direct action (or attention) to a specific portion of the current input.

In a third simulation (Figure 6.3d), a single object is presented and the organisms learn to push the object in response to one signal and to pull the object in response to another signal, whatever the object. These signals

Box 6.3 *(continued)*

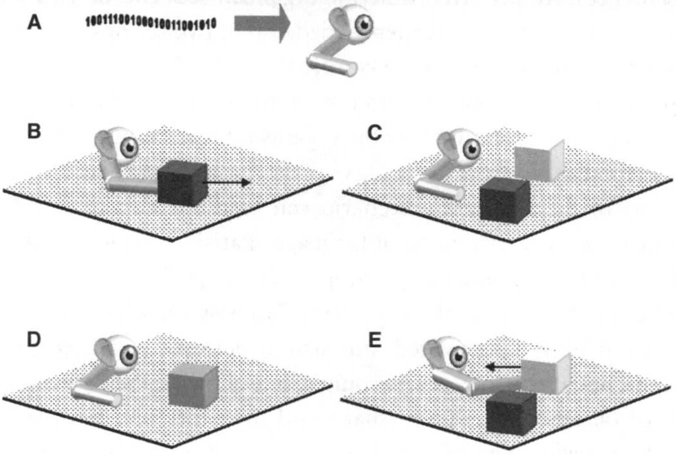

Fig. 6.3 The language model. (A) As in the reaching model, a binary string encodes the characteristics of each organism; in addition to one eye and arm, each organism also has an 'ear', which receives auditory inputs. (B) The organism learns to push objects of one category and pull objects of another category, which are presented at different times. (C) The organism learns to reach one object or another when presented at the same time. (D) The organism learns to push or pull a single object. (E) The organism learns to push one object or pull another when both are present. (See Box 6.3 for details.)

can be interpreted as proto-verbs if verbs are defined as linguistic signals that co-vary with the action to be executed independently of the object on which the action is executed.

Finally, in a fourth simulation (Figure 6.3e) two objects and two signals are presented at the same time and the organisms learn to push or pull one of the two objects in response to the first signal (verb) and they execute the action on one or the other object in response to the second signal (noun). These sequences of two signals can be interpreted as proto-sentences made up of one verb and one noun.

We can also define verbs if we imagine a variation of our scenario. The organism sees a single object, which may be of any shape, and it hears one of two sounds, V1 and V2. When the organism hears V1, it is rewarded if it pushes the object away from itself, and when it hears V2 it is rewarded if it pulls the object toward itself. The sounds now are verbs, not nouns, where a verb is defined as a sound which co-varies with the particular action which must be executed on an object, whatever the object.

If the two scenarios can be put together, then we will see the emergence of simple sentences. At any given time the organism sees two objects that have different shapes, and hears a sequence made up of two sounds. The organism can hear one of four possible sequences: N1+V1, N1+V2, N2+V1, N2+V2. The N sounds allow the organism to select the object on which the action must be executed, while the V sounds allow the organism to select the action that must be executed on the object.

Variations of the simulation scenario can easily be constructed to study the emergence of further aspects of language (Parisi, Cangelosi, and Falcetta, 2002). Imagine that a multiplicity of objects varying in both shape and colour are simultaneously seen by the organism. To know on which of the objects a given action must be executed, our neural network will need not just a single sound but a sequence of two sounds, one allowing the network to select the class of objects with a given shape and the other one for selecting the object with a given color within that class. The sequence of two sounds is a noun phrase. The former sound is a noun because it can occur alone and because in language nouns tend to co-vary with the shape of objects, not with their other properties (Landau and Smith, 1988). The latter sound is a noun modifier, more precisely, an adjective. Noun modifiers can be locatives if the sound accompanying the noun co-varies with the location of the object in the visual field or with respect to another object, not with other, more permanent, properties of objects such as their colour.

Language emerges at the end of the first year of life when infants begin to notice how specific sounds co-vary with specific aspects of their experience, i.e., of what they perceive and what they and other individuals do. Prior to that, during their entire first year, they have learned both how to map non-linguistic inputs into non-linguistic outputs (sensory-motor skills) and how to map heard sounds—which may be produced by others or self-produced—into produced sounds (imitation of sounds, including self-imitation). One can hypothesize that at least two functionally separate networks exist in the brain for doing these two different things for the entire first year of life. At around 1 year, the two networks begin to become connected together. This means that non-linguistic input (e.g., an object is seen) may be mapped into linguistic output (a sound is produced) and linguistic input (a sound is heard) may be mapped into non-linguistic output (an action is executed). This is the beginning of, respectively, language production and language comprehension.

As suggested by our simulations, one may also hypothesize that for some time after the onset of language at around one year the linguistic signals that are produced and understood by children cannot be assigned to distinct classes from the children's point of view, i.e., in their brain, because infants have not had sufficient linguistic experience to notice systematic differences in the co-variation patterns of different classes of signals. On the other hand, many of the early linguistic signals produced by children tend to be recognized and interpreted by adults as nouns (i.e., the names of things). This suggests that 'noun' as a linguistic category may have a developmental advantage over other categories of words. Indeed, our simulations can be interpreted as indicating that nouns—and noun phrases—are, basically, means for directing the attention of the hearer to specific portions of what is currently perceived by the hearer—or *should* be perceived, when the hearer is asked to search for objects with given characteristics. A crucial skill that infants are developing at this age is the ability to selectively attend to portions of what they perceive and, especially, to sharing selective attention with others (e.g., Butterworth and Jarret, 1991). We believe that nouns are well-suited for playing an important role in this process.

Summary

In this chapter, we have introduced Connectionism in an Artificial Life Perspective (CALP). The two key elements of our approach are (1) the use of neural networks, and (2) autonomous organisms that learn, develop, and evolve through Artificial Life. In combination, these elements not only highlight the interactive nature of developmental processes, but also provide a broad, flexible framework for simulating and investigating situated and embodied knowledge. We have also argued, more generally, that by conceptualizing the individual organism as an autonomous agent, complex interactions that span multiple spatiotemporal scales can be systematically studied. One of the valuable benefits of this approach is that the influences of learning, maturation, development, and evolution, often considered in isolation, can all be incorporated into the same modelling process.

To situate our work within a larger perspective, we now raise and respond to what we view as four fundamental issues that can be used to evaluate CALP. First, which features of the embodied and situated organism may constrain the development of representations?

Although a variety of embodied constraints play a role in each of the models that we described (i.e., motor, cognitive, and language), the tracking model is particularly well-suited to address the development of representations. Specifically, this work questions the assumption that prospective, object-oriented activity in infants requires an internal model of the external world. Therefore, rather than building in structures that explicitly support memory-based or predictive processes, our key design strategy is instead to build a sensorimotor (i.e., oculomotor) system in which sensory feedback informs each action. Although the tracking model 'lives' in a snapshot world, it quickly learns to generate actions which are implicitly predictive. In particular, it efficiently tracks moving objects, adjusting its gaze as necessary to keep the target object within view. In addition, we also note that during occlusion, the model shifts its gaze to the location where the target will reappear, although it is not explicitly taught or instructed to do so (see Box 6.2).

How are these future-oriented behaviours achieved without some kind of internal model? There are at least three levels of constraint that operate in the tracking model. First, the model 'selects' its own input. As we highlight elsewhere (e.g., Schlesinger and Parisi, 2001b), we believe that sensory self-selection is a crucial element of development, and more specifically, that it shifts the question of representation development away from 'how an internal copy of the environment is learned', and instead toward 'how an adaptive sensorimotor mapping of the environment is learned'. It is this mapping, we propose, that implicitly encodes key features of the environment. Second, the physical design of our one-eyed organism incorporates a number of complementary constraints, including in particular two visual processing streams, analogous to the fovea and visual periphery. Finally, the specific learning algorithm employed (i.e., reinforcement learning) also constrains the process of representation development, or perhaps more accurately, *enhances the likelihood* that the tracking model will exploit sensorimotor strategies of representation, by shaping the model's performance with evaluation (i.e., a global assessment of behaviour as a whole) rather than by specific, instructive feedback (i.e., by supervised learning).

This suggests a second, fundamental issue: in general, how are embodied and situated constraints implemented in CALP? As we have stressed repeatedly in this chapter, each level of 'physicality' provides a specific set of constraints on development. One level is the morphological design of the organism, which allows it to recruit and exploit geometric features or regularities in the environment. Corresponding to this design, of course, is the structure of the environment, which affords various opportunities for learning and development. For example, the reaching model learns that only two of its three joint DOF are necessary for reaching along a two-dimensional, planar surface. Another level of constraint is imposed by the architecture of the neural network, which in turn is constrained by the design and implementation, where appropriate, of the genetic code that determines the internal and external structure of the organism.

A third fundamental issue concerns the major lessons that CALP may offer for developmental researchers. One implication of our modelling work is a renewed emphasis on the role of action during development, in contrast to recent alternative approaches that have focused on passive, statistical-learning processes (e.g., Marcus, Vijayan, Rao, andVishton, 1999; Saffran, 2003). Note that action plays a particularly important role in our language work, where goal-directed actions are acquired in the context of a linguistic signal. We hypothesize that it is the coordination of these actions with ongoing linguistic input that enables the infant to construct implicit categories for 'verb' and 'noun', and to begin recognizing and ultimately using these proto-words in conventional ways.

Another implication of CALP is the notion of an 'emergent constraint'. This is the idea that the outcome of a developmental process need not be programmed in by maturation, but instead may occur as the result of successive learning experiences that the organism determines or selects for itself. For example, in our reaching model, the upperarm and forearm segments are controlled by parallel, analogous systems, and yet the model consistently learns to efficiently move the upper arm first. This temporal bias suggests a proximodistal developmental pattern, as the organism first learns to control limb segments that are closest to the central body axis or torso. Therefore, the proximodistal pattern constrains future developmental experiences, without being explicitly pre-programmed into development itself.

Finally, what concrete predictions, if any, do our CALP models suggest? First, with respect to motor development, our reaching model predicts a proximodistal pattern of muscle control, specifically beginning first in the shoulder and upperarm muscles, followed by the forearm muscles. The reaching model also predicts that while wearing displacing prisms, infants should learn to adapt or recalibrate their reaches toward real, three-dimensional objects before they adapt their reaches toward virtual objects.

Second, the tracking model predicts that in preferential-looking studies that implement the violation-of-expectation paradigm, infants should generate different gaze patterns during possible and impossible events. In particular, the model predicts that infants' gaze patterns during the habituation (or familiarization) event and the possible event should be most similar, while the impossible event should provoke largely novel scanning or gaze patterns.

Third, one of the predictions of the language model is that separate neural structures for (1) mapping non-linguistic, sensorimotor input onto sensorimotor output, on the one hand, and (2) mapping linguistic, auditory input onto linguistic, vocal output, on the other hand, initially develop along parallel, but independent trajectories. The model also predicts that these structures or pathways become coordinated (e.g., directly or indirectly connected) at roughly the end of the first year of life, enabling the infant to use their existing sensorimotor knowledge to begin inferring meaning or intent from linguistic signals.

In conclusion, despite the many unique advantages of studying neural networks in an Artificial Life perspective, we acknowledge that our approach is not only one of many different modelling techniques, but more importantly, that it also faces a number of significant challenges (e.g., Parisi and Schlesinger, 2002; Schlesinger and Parisi, 2001b). For example, how can an organism that is capable of only sensorimotor representation ultimately acquire the capacity for symbolic mental activity (e.g., metaphor, deduction, counterfactual reasoning, etc.)? While there are a variety of theories about how abstract mental symbols may develop out of embodied knowledge (e.g., Johnson, 1987; Piaget, 1952), it is certainly a more difficult task to demonstrate this process convincingly through simulation. Nevertheless, we are confident that the CALP will be able to address these kinds of questions, and more generally, continue to play a vital role in developmental research.

Acknowledgements

This work was supported in part by a grant to the first author from the National Institute of Child Health and Human Development (1R03 HD40780–02).

Glossary

DOF problem Systems with multiple free parameters are often underconstrained: they have more degrees of freedom (DOF) than are necessary to solve a given task. During arm movements, for example, humans require only three DOF (i.e., moveable limb segments) to reach for or point to any object in a three-dimensional workspace. In this case, the DOF problem arises because of redundant or unnecessary limb segments and their corresponding joints (waist, shoulder, elbow, wrist, etc.), which create an unlimited number of possible movement paths.

Genetic epistemology A formal study of cognitive development, made famous by Piaget, that defines knowledge acquisition as a constructive process influenced by both genetic and experiential factors.

Genome The entire pool of genetic material in a population of organisms.

Genotype The set of genetic material inherited by an individual. Genotype also refers to specific portions of genetic material (e.g., a gene) that encode a particular characteristic (e.g., the genotype for brown eyes).

Ontogeny The process of change during the lifespan (i.e., from conception to death) of an individual organism.

Phenotype The observable characteristics of an individual, determined in part by their genotype.

Phylogeny The process of change that spans multiple generations (i.e., evolution).

References

Aslin, RN (1987). Visual and auditory development in infancy. In JD Osofsky (ed.) *Handbook of infant development*, 2nd edn, pp. 5–97). New York: Wiley and Sons.

Baillargeon, R (1986). Representing the existence and the location of hidden objects: Object permanence in 6- and 8-month-old infants. *Cognition*, 23, 21–41.

Baillargeon, R (1999). Young infants' expectations about hidden objects: A reply to three challenges. *Developmental Science*, 2, 115–132.

Baillargeon, R and DeVos, J (1991). Object permanence in young infants: Further evidence. *Child Development*, 62, 1227–1246.

Ballard, DH, Hayhoe, MM, Pook, PK, and Rao, RPN (1997). Deictic codes for the embodiment of cognition. *Behavioral and Brain Sciences*, 20, 723–767.

Bernstein, NA (1967). *The coordination and regulation of movements.* Oxford: Pergamon.

Bertenthal, BI, Campos, JJ, and Kermoian, R (1994). An epigenetic perspective on the development of self-produced locomotion and its consequences. *Current Directions in Psychological Science*, 3, 140–145.

Berthier, NE (1996). Learning to reach: A mathematical model. *Developmental Psychology*, 32, 811–823.

Berthier, NE, Bertenthal, BI, Seaks, JD, Sylvia, MR, Johnson, RL, and Clifton, RK (2001). Using object knowledge in visual tracking and reaching. *Infancy*, 2, 257–284.

Bremner, JG (1985). Object tracking and search in infancy: A review of data and a theoretical evaluation. *Developmental Review*, 5, 371–396.

Brooks, RA (1991). Intelligence without representation. *Artificial Intelligence*, 47, 139–159.

Butterworth, G and Jarret, N (1991). What minds have in common in space: Spatial mechanisms serving joint visual attention in infancy. *British Journal of Developmental Psychology*, 9, 55–72.

Cangelosi, A and Parisi, D (1998). The emergence of 'language' in an evolving population of neural networks. *Connection Science*, 10, 83–97.

Clifton, RK, Muir, DW, Ashmead, DH, and Clarkson, MG (1993). Is visually guided reaching in early infancy a myth? *Child Development*, 64, 1099–1110.

Clifton, RK, Rochat, P, Robin, DJ, and Berthier, NE (1994). Multimodal perception in the control of infant reaching. *Journal of Experimental Psychology: Human Perception and Performance*, 20, 876–886.

Elman, J, Bates, E, Karmiloff-Smith, A, Johnson, M, Parisi, D, and Plunkett, K (1996). *Rethinking innateness: A connectionist perspective on development.* Cambridge, MA: MIT Press.

Georgopolous, AP, Schwartz, AB, and Kettner, RE (1986). Neuronal population coding of movement direction. *Science*, 233, 1416–1419.

Gibson, EJ (2000). Perceptual learning in development: Some basic concepts. *Ecological Psychology*, 12, 295–302.

Haith, MM (1998). Who put the cog in infant cognition? Is rich interpretation too costly? *Infant Behavior and Development*, 21, 167–179.

Held, R and Hein, A (1963). Movement produced stimulation in the development of visually guided behavior. *Journal of Comparative and Physiological Psychology*, 56, 872–876.

Hinton, GE and Nolan, SJ (1987). How learning can guide evolution. *Complex Systems*, 1, 495–502.

Holland, JH (1975). *Adaptation in natural and artificial systems*. Ann Arbor, MI: University of Michigan Press.

Johnson, M (1987). *The body in the mind: The bodily basis of imagination, reason, and meaning*. Chicago, IL: University of Chicago Press.

Johnson, SP, Amso, D, and Slemmer, JA (2003). Development of object concepts in infancy: Evidence for early learning in an eye tracking paradigm. *Proceedings of the National Academy of Sciences, USA, 100*, 10568–10573.

Jones, G, Ritter, FE, and Wood, DJ (2000). Using a cognitive architecture to examine what develops. *Psychological Science*, 11, 93–100.

Landau, B and Smith, LB (1988). The importance of shape in early lexical learning. *Cognitive Development*, 2, 291–321.

Marcus, GF, Vijayan, S, Rao, SB, and Vishton, PM (1999). Rule learning by seven-month-old infants. *Science*, 283, 77–80.

Mareschal, D, Plunkett, K, and Harris, P (1999). A computational and neuropsychological account of object-oriented behaviours in infancy. *Developmental Science*, 2, 306–317.

McDonnell, PM and Abraham, WC (1979). Adaptation to displacing prisms in human infants. *Perception*, 8, 175–185.

Miglino, O, Nolfi, S, and Parisi, D (1996). Discontinuity in evolution: How different levels of organization imply pre-adaptation. In RK Belew and M Mitchell (eds.) *Adaptive individuals in evolving populations*, pp. 399–415. Reading, MA: Addison-Wesley.

Munakata, Y, McClelland, JL, Johnson, MH, and Siegler, RS (1997). Rethinking infant knowledge: Toward an adaptive process account of successes and failures in object permanence tasks. *Psychological Review*, 104, 686–713.

Nolfi, S and Parisi, D (1993). Self-selection of input stimuli for improving performance. In GA Bekey and KY Goldberg (eds.) *Neural networks in robotics*, pp. 403–418. Boston, MA: Kluwer.

Nolfi, S and Parisi, D (2002). Evolution of artificial neural networks. In MA Arbib (ed.) *Handbook of brain theory and neural networks*, pp. 418–421. Cambridge, MA: MIT Press.

Nolfi, S, Floreano, D, Miglino, O, and Mondada F (1994). How to evolve autonomous robots: Different approaches in evolutionary robotics. In R Brooks and P Maes (eds.) *Proceedings of the international conference artificial life IV*, pp. 190–197. Cambridge, MA: MIT Press.

O'Regan, JK (1992). Solving the 'real' mysteries of visual perception: The world as an outside memory. *Canadian Journal of Psychology*, 46, 461–488.

O'Regan, JK and Noe, A (2001). A sensorimotor account of vision and visual consciousness. *Behavioral and Brain Sciences*, 24, 939–1031.

Parisi, D (1996). Computational models of developmental mechanisms. In R Gelman and TK Au (eds.) *Perceptual and cognitive development*, pp. 373–412. San Diego, CA: Academic Press.

Parisi, D (1997). Active sampling in evolving neural networks. *Human Development*, 40, 320–324.

Parisi, D and Schlesinger, M (2002). Artificial life and Piaget. *Cognitive Development*, 17, 1301–1321.

Parisi, D, Cangelosi, A, and Falcetta, I (2002). Verbs, nouns, and simulated language games. *Italian Journal of Linguistics*, 14, 99–114.

Parisi, D, Cecconi, F, and Nolfi, S (1990). Econets: Neural networks that learn in an environment. *Network*, 1, 149–16.

Piaget, J (1952). *The origins of intelligence in children*. New York: International Universities Press.

Saffran, J (2003). Statistical language learning: Mechanisms and constraints. *Current Directions in Psychological Science*, 12, 110–114.

Schlesinger, M (2003). A lesson from robotics: Modeling infants as autonomous agents. *Adaptive Behavior*, 11, 97–107.

Schlesinger, M (2004). Evolving agents as a metaphor for the developing child. *Developmental Science*.

Schlesinger, M and Barto, A (1999). Optimal control methods for simulating the perception of causality in young infants. In M Hahn and SC Stoness (eds.) *Proceedings of the twenty first annual conference of the Cognitive Science Society*, pp. 625–630. Mahwah, NJ: Erlbaum.

Schlesinger, M and Casey, P (2003). Visual expectations in infants: Evaluating the gaze-direction model. In CG Prince, L Berthouze, H Kozima, D Bullock, G Stojanov and C Balkenius (eds.) *Proceedings of the third international workshop on epigenetic robotics: Modeling cognitive development in robotic systems*, pp. 115–122. Sweden: Lund University Cognitive Studies.

Schlesinger, M and Parisi, D (2001a). Multimodal control of reaching: The role of tactile feedback. *IEEE Transactions on Evolutionary Computation: Special Section on Evolutionary Computation and Cognitive Science*, 5, 122–128.

Schlesinger, M and Parisi, D (2001b). The agent-based approach: A new direction for computational models of development. *Developmental Review*, 21, 121–146.

Schlesinger, M and Parisi, D (eds.) (2004). Beyond backprop: Emerging trends in connectionist models of development. [Special section]. *Developmental Science*.

Schlesinger, M and Young, ME (2003). Examining the role of prediction in infants' physical knowledge. In R Alterman and D Kirsh (eds.), *Proceedings of the Twenty-Fifth Annual Conference of the Cognitive Science Society*, pp. 1047–1052. Boston: Cognitive Science Society.

Schlesinger, M, Parisi, D, and Langer, J (2000). Learning to reach by constraining the movement search space. *Developmental Science*, 3, 67–80.

Smith, LB (1999). Do infants possess innate knowledge structures? The con side. *Developmental Science*, 2, 133–144.

Spelke, ES (1985). Preferential looking methods as a tool for the study of cognition in infancy. In G Gottlieb and N Krasnegor (eds.) *Measurement of audition and vision in the first year of postnatal life*, pp. 323–363. Norwood, NJ: Ablex.

Spelke, ES (1998). Nativism, empiricism, and the origins of knowledge. *Infant Behavior and Development*, 21, 181–200.

Thelen, E and Smith, LB (1994). *A dynamic systems approach to the development of cognition and action*. Cambridge, MA: MIT Press.

Thelen, E, Corbetta, D, Kamm, K, Spencer, JP, Schneider, K, and Zernicke, RF (1993). The transition to reaching: Mapping intention and intrinsic dynamics. *Child Development*, 64, 1058–1098.

Thelen, E, Schöner, G, Scheier, C, and Smith, LB (2001). The dynamics of embodiment: A field theory of infant perseverative reaching. *Behavioral and Brain Sciences*, 24, 1–34.

Vereijken, B, Emmerik, REA van, Whiting, HTA, and Newell, KM (1992). Free(z)ing degrees of freedom in skill acquisition. *Journal of Motor Behavior*, 24, 133–142.

Von Hofsten, C (1984). Developmental changes in the organization of prereaching movements. *Developmental Psychology*, 20, 378–388.

Chapter 7

Using robots to study the mechanisms of imitation

Yiannis Demiris

Introduction

In recent years we have seen a large increase in the number of research projects that take inspiration from the life sciences to design mechanisms that allow robots to imitate other robots or humans (see reviews by Breazeal and Scassellati, 2002; Schaal, 1999). These mechanisms allow robots to take advantage of knowledge in their social environment to improve their abilities to solve tasks. The engineering reasons for looking at biological systems for inspiration are clear: these systems have evolved excellent imitation capabilities, and are routinely using them for their benefit (Dautenhahn and Nehaniv, 2002; Meltzoff and Prinz, 2002). I would like, however, to concentrate on another issue, which, although rarely stated as an explicit goal is frequently implicit in much of the biologically inspired research. That is, we would like our robot experiments to feed some knowledge back to the life sciences (Demiris and Hayes, 2002; Webb, 2001).

It is unclear whether we are achieving this goal when it comes to imitation. Why might this be? There are obvious technological reasons: the sophistication of the robots currently available leaves much to be desired. Although there have been great advances in robot vision and motor systems (see Box 7.1 for a brief survey of the tools currently available for researchers interested in using robots for modelling experiments), we are still a long way from the functionality and versatility of their human equivalents.

There are also methodological issues. Most experimenters are content when the robotic system displays a behaviour comparable (at some level of abstraction) to that of its biological counterpart. I will argue that if robotics research is to deliver any significant feedback, it should aim to produce *plausible explanations* about the behaviour of the modelled system, and most importantly it should aim to generate *testable predictions*.

To elaborate on these points, I will present three case studies in which robots were used as models of imitation mechanisms, in an increasingly detailed way.

1. In the first example (Hayes and Demiris, 1994), we investigated how knowledge can be passed from one robot to another, in experiments where the task was to negotiate the corners of a maze. The focus of these experiments was primarily to investigate the advantages (for robots) of using imitation, and the comparison with biological systems (matched dependent behaviour) was an additional dimension.
2. In the second example (Demiris and Hayes, 1996; Demiris *et al.*, 1997), we designed a computational architecture, and implemented it in a robot stereo head, as an explicit attempt to model the Active Intermodal Matching (AIM) hypothesis put forward by Meltzoff and Moore (1997). The model was exploratory in nature, investigating:
   - What are the *necessary* structures that need to be in place for the AIM mechanism to work, and
   - Whether we can build computational equivalents that use information known to be available in biological brains, such as the postures of the demonstrator, perceived visually, and those of the imitator, perceived proprioceptively (Perrett *et al.*, 1991; Tillery *et al.*, 1996; Scott and Kalaska, 1995).
3. In the third example (Demiris, 1999; Demiris and Hayes, 2002; Demiris, 2002), simulated robots were used to examine the behaviour of mirror neurons (Gallese *et al.*, 1996). Apart from suggesting possible explanations for the available data on mirror neurons, we explicitly tried to generate testable predictions by manipulating the model's parameters to derive model data for which, currently, there are no equivalent biological data.

Learning to navigate through imitation

In this first study, we investigated the mechanisms underlying the ability to learn by imitation. One robot was used as a demonstrator in a task that involved navigation through a maze. A second robot acted as a learner; it was equipped with a basic imitation mechanism which consisted of maintaining a constant distance between the demonstrator and the imitator, essentially following the demonstrator (see Hayes and Demiris, 1994, for technical details). Figure 7.1 shows a snapshot of the scenario, while Figure 7.2 gives a top view of the learning architecture as it was used in these experiments.

Over the course of the demonstration, the imitator was associating the environmental configuration with the actions it was performing because it was imitating (following) the demonstrator. When the demonstration is finished,

Fig. 7.1 A snapshot of the experimental set-up.

the imitator's learned knowledge is evaluated by placing the imitator in the maze and observing its behaviour.

These experiments served as an important demonstration of the advantages of using imitation to learn what is needed to perform a task and when. However, from a modelling point of view the resulting data were less interesting. The model we were aiming for when we were doing these experiments

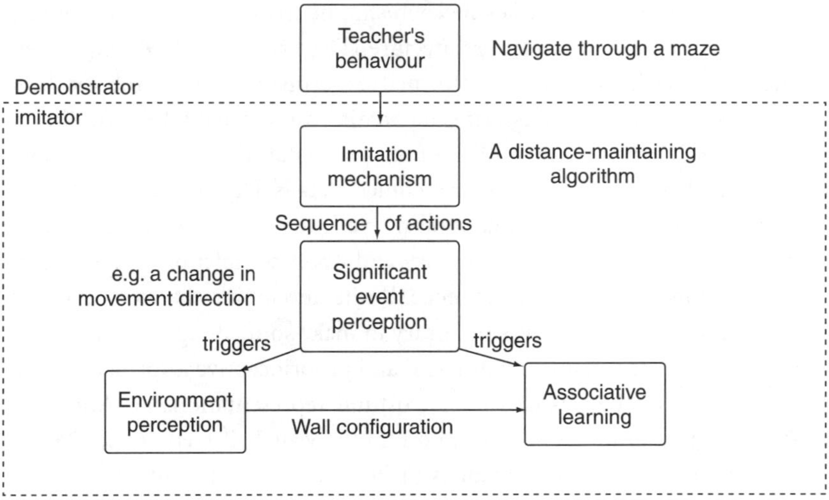

Fig. 7.2 The outline of the learning architecture: the learner is following the teacher, associating significant events (such as changing direction) with the environmental configuration, learning 'what to do and when'.

was a social learning behaviour known as 'matched dependent behaviour' (Miller and Dollard, 1941). This behaviour occurs when rats negotiating a maze receive reinforcement if they match the actions of a leader rat. However because information about the rat's behaviour and physiology was not part of the design of the system, we could not formulate a hypothesis that would be detailed enough to lend itself to a robotic experiment.

This led us to the first principle when using robots as explanatory models of behaviour: one should identify a cognitive hypothesis that is detailed enough to be computationally implementable; for that to be possible, the hypothesis should minimally specify the flow of information in the system, as well as the type of operations that are applied to this information at each stage.

Armed with this initial knowledge, we proceeded to examine a more well-defined hypothesis: Andrew Meltzoff's Active Intermodal Matching Hypothesis (Meltzoff and Moore, 1997).

Modelling active intermodal matching

One of the most interesting sets of experimental data regarding infant imitation abilities comes from facial gestures imitation experiments (Meltzoff and Moore, 1977, 1997). Infants of a very young age have been shown to be capable of imitating a range of facial gestures, including mouth opening, tongue protrusion and lip protrusion among others. Meltzoff and Moore (1997) have proposed the Active Intermodal Matching (AIM) process (summarized in Figure 7.3) as the key mechanism underlying this imitation ability. We designed a computational architecture (Demiris *et al.*, 1997) equivalent of the AIM mechanism, and performed experiments with a robotic head (Figure 7.4) in order to investigate the architecture's capabilities and limitations. Figure 7.5 provides an outline of our computational architecture, showing in parentheses the most relevant technical details. During a demonstration of human head movements (e.g. nodding 'yes') the robot extracts a sequence of postural targets, stores it in memory and upon completion of the human demonstration, replicates the sequence. While developing the computational implementation of AIM, it was necessary to make some design decisions. The most important decision was that visual and proprioceptive information were to be encoded at the postural level. Postural representations are known to exist in the human/monkey brains, both in the visual (Perrett *et al.*, 1991) as well as the proprioceptive modalities (Tillery *et al.*, 1996; Scott and Kalaska, 1995).

Our attempt to build a model of imitation and implement it within a robot head raised a number of new empirical questions. In particular, there

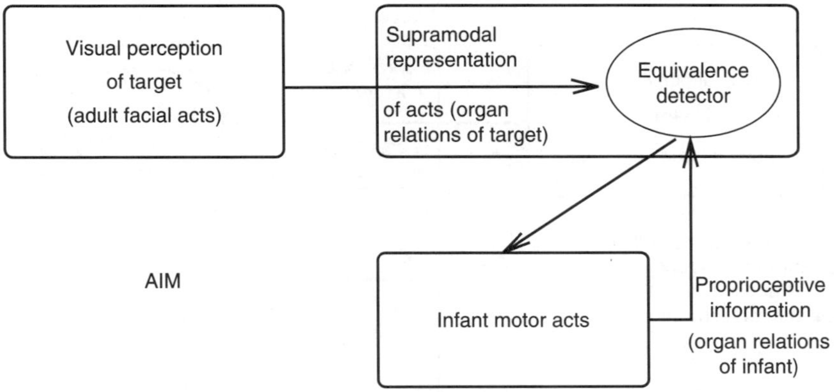

Fig. 7.3 The AIM hypothesis (adapted from Meltzoff and Moore, 1997): the movement of the demonstrator is encoded in a supramodal representation (a set of 'organ relations', for joint angles, distances between limbs, among others), which is subsequently reproduced by the infant; the infants monitors its own organ relations through proprioception, and outputs the motor commands that will make them equivalent to the demonstrator's.

is insufficient information about the nature of what is termed 'supramodal representation' in the AIM hypothesis. In our implementation this took the form of geometric relations (postural configurations), but this was an engineering decision rather than one based on experimental data. Furthermore, in the case of delayed imitation, the way that the demonstration is stored in

Fig. 7.4 A snapshot of the experimental set-up.

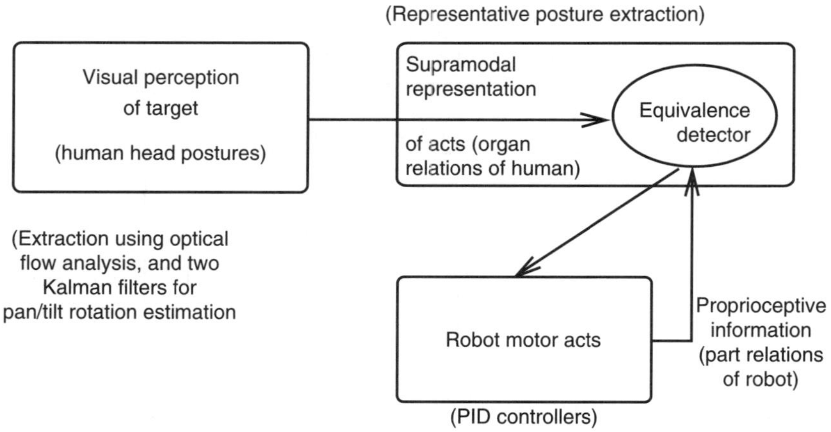

Fig. 7.5 Schema of the computational architecture implementing AIM: the movements of the human demonstrator are represented as a series of postures, which are subsequently reproduced by the robot head.

memory is also not known. In our implementation, we chose not to store the entire sequence of postures in memory, but rather, to filter the sequence and store only the representative postures that are sufficient to describe the postural trajectory. Finally, the nature of the mechanism for comparing inputs from two different modalities (the 'equivalence detector' in AIM) is not well known. Again in our implementation we have used geometric representations, more recently encoding visual and proprioceptive signals using spiking neurons, and making frequency-based comparisons (Eneje and Demiris, 2003), but there is a large number of other possible representations one could use.

These experiments illustrated that developing computational implementations of a hypothetical mechanism strengthens the hypothesis, by providing a computational 'existence proof' that these mechanisms suffice to generate the target behaviours. It also helps to raise a number of new questions that can lead to further psychological experiments. The questions that arise from this study are mainly pointing out parts of mechnanisms that require further specification. It would be better if the computational experiments could formulate *testable predictions*. However, the level of description of the AIM mechanism, as well as that of our computational model, did not lend itself naturally to this. There is a lack of parametrizable and quantifiable mechanisms that can be manipulated to drive the system into situations not tested in biological systems.

Thus we arrive at a second principle of modelling: one should identify in their computational architecture the parametrizable dimensions, then push the parameters to situations for which we do not have experimental data from the life sciences, thus offering testable predictions.

We put this suggestion into action in our third study: modelling mirror neurons.

Motor theories of perception and the modelling of mirror neurons

The third case study was inspired by a range of neurophysiological, psychophysical and brain imaging data that point to a significant involvement of the imitator's motor systems during the perception of an action.

Neurophysiological experiments with macaque monkeys (Gallese *et al.*, 1996; Rizzolatti *et al.*, 1996) have discovered an important class of neurons in the rostral part of inferior area 6 (F5) of macaque monkeys, which were termed 'mirror' neurons. These neurons are active when the monkey performs an action, as well as when the monkeys observes the same action being performed by a demonstrator. Several types of mirror neurons were found: grasping, tearing, manipulating and placing objects neurons, among others. Some of the neurons were active only during the demonstration while some others remained active for a while after the end of the demonstrated action. These data, among others (Carey *et al.*, 1997), suggest that there is a motor system involvement during observation of movement.

In humans, several experiments have investigated the interplay between action generation and action perception. For example (Fadiga *et al.*, 1995), stimulated the motor cortex of human observers and recorded the motor-evoked potentials (MEPs) from hand muscles, utilizing the assumption that if action observation activates the premotor cortex (as it does in monkeys), this activation should induce an increase of the MEPs elicited by the magnetic stimulation of the motor cortex. They found a significant increase of the MEPs when subjects observed movements; additionally the pattern of muscle activation was very similar to the pattern of muscle contraction present during the execution of the same action. That is, the increase was present only in those muscles that are active when the human subjects executed these actions. Brain imaging data from PET and fMRI experiments have also shown activation of motor areas during the observation of actions (Decety *et al.*, 1997; Meltzoff and Decety, 2003). The results of eye-tracking experiments examing the fixation behaviour of humans observing a demonstrator during a task involving the construction of small structures using blocks were also seen as supportive of the motor theory of perception (Flanagan and Johansson, 2003).

The computational architecture

The fact that motor systems are involved during the observation of an action, can be captured through the use of *internal models*. We used two different types of internal models:

1. *Inverse models* (also known as controllers, or behaviours) which, given a goal and the current state of the controlled system, output the necessary motor commands that are needed in order to achieve or maintain that goal. These structures are frequently used in engineering (Narendra and Balakrishnan, 1997) and have been used for modelling motor planning and motor control (Wolpert and Kawato, 1998; Wolpert *et al.*, 2003). The implementation of inverse models varies according to the application (Demiris and Hayes, 2002; Demiris and Johnson, 2003); it is also possible to learn these models through reinforcement learning, where different motor commands are attempted and their result is evaluated with respect to the desired goal. The ones that achieve the goal are reinforced and stored for future use.
2. *Forward Models* (also known as predictors) which, given motor commands and the current state of the controlled system, output the predicted next state of that system (or the predicted feedback (Karniel, 2002). Like inverse models, they have been used in motor control modelling (Miall and Wolpert, 1996). In fact, it has been suggested that the cerebellum acts like a forward model (Miall *et al.*, 1993).

Figure 7.6 shows a typical arrangement of an inverse and a forward model, as used together in control systems.

With this arrangement, an inverse model, encoding a particular behaviour, is given the current state and a target goal (for example, raise arm to 45°). The inverse model outputs the motor commands that would achieve that target goal, and these are sent to the musculoskeletal system, which in turn feeds back proprioceptive information. However, this feedback takes a significant amount of time before it is available, so, in parallel to this, the motor commands are also sent to a forward model which will output a prediction of what the next state will be. This predicted state is compared with the target goal, and any errors are used to adapt the inverse model so that it performs better.

We can introduce parallel processing into this configuration. Multiple pairs of inverse and forward models (Figure 7.7), encoding different behaviours, are sent the target goal. They all propose motor commands that they hypothesize will achieve the target goal. These motor commands are sent to the forward models which output a set of predicted next states. The pair that has generated the next state prediction most compatible with the target goal receives

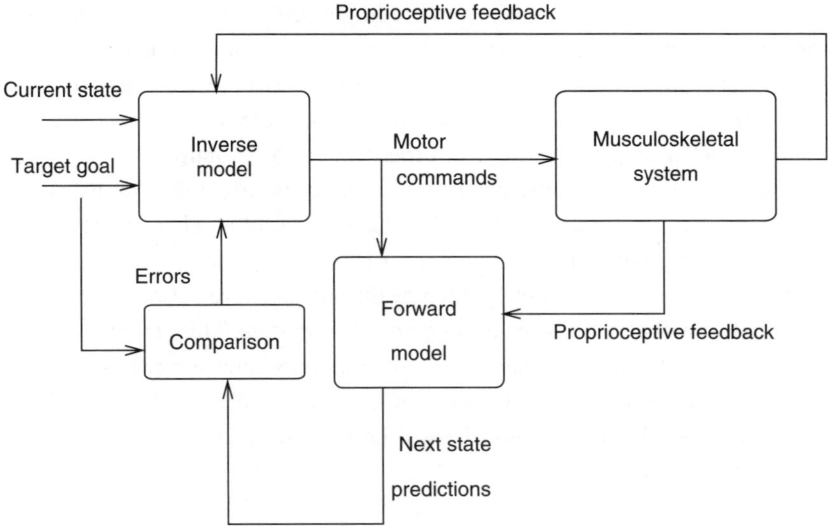

Fig. 7.6 Single action control: the inverse model outputs motor commands that are sent to the musculoskeletal system, as well as a forward model which will predict the next state of the system; this prediction is compared with the desired goal to provide feedback to the inverse model.

reinforcement (increases its confidence), while the rest use the errors to adapt their internal parameters so they can suggest better hypotheses.

We use this parallel architecture in two ways. By feeding in a target goal, the architecture can be used to plan and select the most appropriate motor

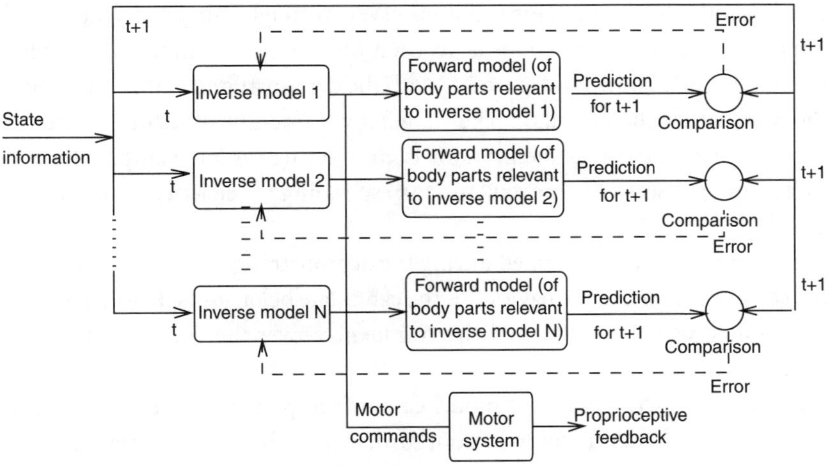

Fig. 7.7 A top view of the computational architecture: several pairs of behaviours and forward models receive state information at time t and predict the state at time t+1; the errors in the prediction are used as feedback to reinforce the behaviours that perform well and increase their confidence.

commands to achieve it. However, by feeding in the state of the *demonstrator* instead, we can also use this architecture to *recognize* what is being demonstrated. The inverse models receive the current state of the demonstrator, and output the motor commands, *as if they were in that state, and had to achieve the respective behaviours they encode.* In other words, the imitator uses its motor systems to simulate mentally possible behaviours (taking the demonstrator's perspective) while the demonstrator's behaviour unfolds. These motor commands are sent to the forward models only (they are inhibited from being sent to the musculoskeletal system) which again output a prediction of the next state, but now that state is *the demonstrator's next state.* This state prediction is compared with the next state of the demonstrator, and the errors are used as reinforcement signals to the various inverse models. At the end of the demonstration, the inverse model with the highest confidence is selected as the behaviour to be imitated.

The experiments

We implemented this architecture using two simulated humanoid robots. The simulation was real-physics-based (for the technical details see, Demiris, 1999; Demiris and Hayes, 2002 and Box 7.1), so movements were subject to gravity and friction. The demonstrator robot executed one of several possible behaviours (moving the arms to achieve the different letters of the International Standard Semaphore Code (ISSC) from the rest position). While the demonstration unfolds, the imitator robot observes (through simulated vision) and internally generates predictions as to what the next state will be. The experimental set-up is shown in Figure 7.8, with the demonstrator in the top left and the imitator on the top right. The six windows underneath demonstrate the current possibilities that the imitator is considering (as an example, the first three windows show the movements corresponding to the letters L, K, I of the ISSC).

The errors that are generated during the demonstration are used to increase the confidence for each behaviour of the candidate behaviours. Figure 7.9 gives an example of the evolution of the confidences over the course of a typical experiment.

I will now describe how the model can be interpreted as an explanation of the mechanism underlying imitation; furthermore, by hypothesizing that the inverse models are the computational equivalent of the mirror neurons, the model also makes testable predictions regarding the behaviour of the mirror neurons.

Fig. 7.8 The experimental set-up comprising of two dynamically simulated humanoids (demonstrator shown in the top left panel, and the imitator on the top right panel; the lower six panels show the current possibilities that the imitator is considering).

Explanations by the model

The explanation offered by our computational model for the biological experimental data described earlier, is that the motor system is activated in order to generate and internally simulate candidate behaviours, as well as generating predictions regarding the incoming perceptual data from the demonstrator. In other words, an action is effectively understood by internally generating it. The observer imitates the demonstrated movement internally, even when it does not do so externally. This feature of the architecture explains why physically imitating a set of demonstrated movements does not aid their later recall (Zimmer and Engelkamp, 1996), as well as why physical rehearsal of a demonstrated behaviour does not lead to any significant differences in the levels of performance improvement from mental rehearsal or mere observation (Vogt, 1995). Since observation, imagery and imitation are done using largely the same structures (behavioural modules and forward models), the same laws should govern their operation. This explanation is also consistent with Decety's

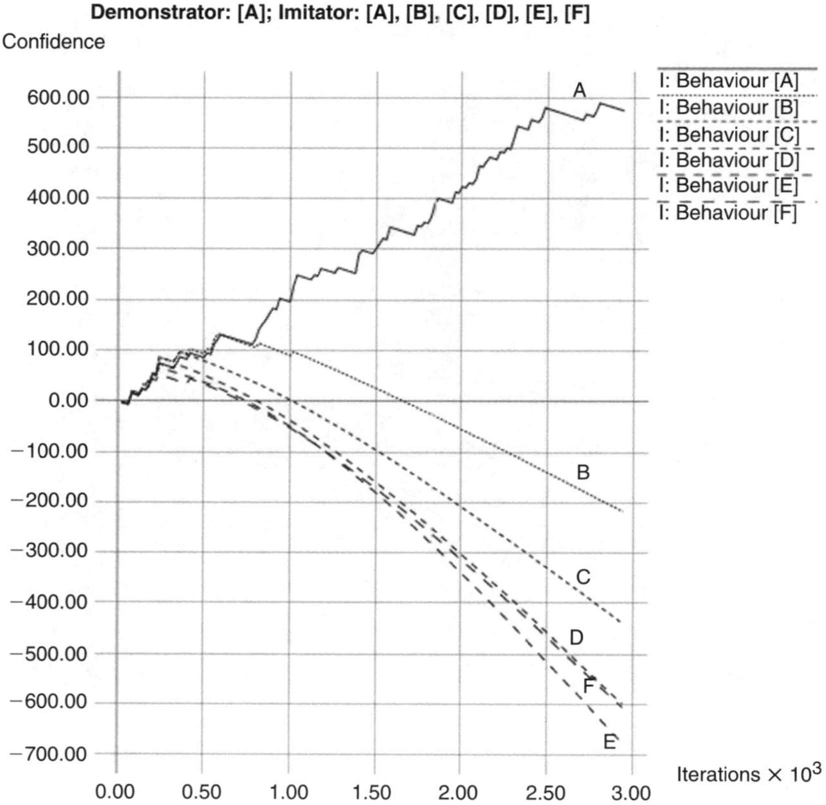

Fig. 7.9 The evolution of confidences in a typical experiment: the demonstrator performs movement (A); the corresponding movement for the imitator is properly activated, while the others receive negative reinforcement.

mental chronometry data indicating that it takes roughly the same time to perform a task mentally or physically (Decety, 1996).

Model predictions

Additional interesting results came from attempting to generate testable predictions. We performed experiments altering the model and scenario parameters to investigate conditions that had not been investigated in the biological side. More specifically, we wanted to investigate the effects of altering the speed of the demonstration. To achieve this, we altered the dynamical properties of the two robots (weight and friction) in a systematic way (Demiris, 1999). We discovered that there are situations where the imitator does not correctly recognize the demonstrated movement despite the fact that it is included in the imitator's repertoire. Figure 7.10 illustrates this. Here, the imitator is

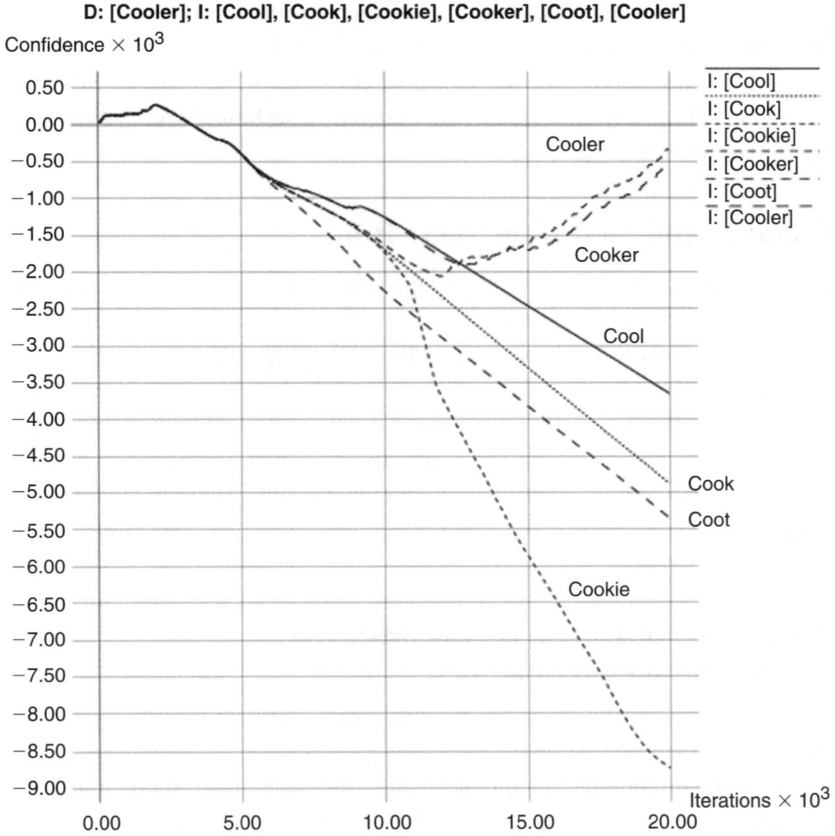

Fig. 7.10 The imitator has the demonstrated behaviour in its repertoire but it cannot be recognized because of the speed of the demonstration.

400 per cent heavier than the demonstrator, and although the demonstrated behaviour (a series of arm movements corresponding to ISSC 'Cooler') is in the imitator's repertoire, its confidence ends up very low, because the behaviour cannot be executed by the imitator at high speeds.

If we map this onto the behaviour of mirror neurons, we arrive at the following testable prediction: a mirror neuron that is active during an action demonstration will not be active (or be less active), if the action is performed at speeds that the monkey cannot attain.

Furthermore, the experimental data in Figure 7.9 show that at the start of the demonstration there are more behaviours with a positive level of confidence as compared to later in the demonstration, when its content has been clarified. By mapping the confidence levels to neuronal activity levels we can predict that there will be a higher level of brain activity at the start of an observation, and that it will decrease as the demonstration unfolds.

It is important to note that both the predictions above are testable. However, to the best of our knowledge, there are currently no biological data available for the scenarios they describe.

Robotics and neuroconstructivism

In the experiments above, the unifying scheme was that of modelling imitation mechanisms using robots. Various functional constraints extracted from biological data were used when designing the computational model. In the following sections, I will make explicit some of the key decisions taken, focusing on the following unifying questions of this book: (a) what functional brain, embodiment, and situatedness constraints operate on the representation development in our models? (b) How are they implemented in the models? (c) What have we learned from the modelling exercises? (d) Do these models make concrete predictions?

Constraints operating on the process of representation development

What functional brain, embodiment and situatedness constraints operate on the process of representation development in the three models described previously? In the first model, the intention was to illustrate learning by imitation; by imitating the demonstrator, the imitator robot learned simple knowledge representations, i.e. response actions to particular environmental configurations. Based on the knowledge we have for the capabilities of rats, this is within the learning capacities of the modelled system. However, no specific rat functional brain constraints were used in the design of the model. Its embodiment also makes limited contribution to its representation development: there is a localized view of the environment (no global maps are present or learned), and the imitator only picks features of the environment that are perceivable through its sensors.

In the subsequent two systems, the functional brain constraints present are more significant. In the head movement imitation experiments, we used only representations known to be present in the primate brain: state representations of the demonstrator's movement are encoded and imitated as postural sequences; there are good evidence for such representations both in the visual (Perrett *et al.*, 1991) as well as the proprioceptive modalities (Tillery *et al.*, 1996; Scott and Kalaska, 1995). We have implemented these constraints by encoding postures either as sets of angles (Demiris *et al.*, 1997), or spiking neuron activations (Eneje and Demiris, 2003). In the mirror neuron modelling experiments, apart from the constraints above regarding the state representations, we added an additional one, inspired by the information processing parallelism observed in the primate brain, and in particular in the motor system (Rizzolatti *et al.*, 1998): the nature of motor representations must be distributed, having no centralized supervising controller. As we saw earlier, we have implemented this constraint by utilizing multiple inverse and forward models (Wolpert and Kawato, 1998; Demiris and Hayes, 2002), and relying on mutual inhibition for achieving the selection of an appropriate inverse model. The embodiment constraints have a more important effect on the representation development: due to the dual use of the motor system, both as an execution system as well as a mental simulation and perception system, any changes that affect the inverse models (through learning new control programs, for example new ways for controlling an arm) will also have an impact on the range of behaviours that the system can perceive. In other words, there is co-development process between the motor system and the system that perceives actions of others.

Concrete predictions of the models

The first two models were used mostly as tools for developing new empirical questions, and as such did not produce any concrete predictions. However, the third model delivered a concrete and testable prediction: mirror neurons in the macaque premotor cortex should be sensitive to the velocity of the movement of the demonstrator; they should exhibit little or no activation when the movement is demonstrated at speeds that are unattainable by the monkey itself. This is an effect of the tight coupling between action and perception systems: the motor systems are activated to mentally simulate the demonstrated actions and provide predictions for the forthcoming states of the demonstration.

Concluding remarks—what have we learned?

In this chapter, I described my work on modelling imitation mechanisms using robots. My experiments have proceeded from trying to model high-level behaviours such as matched dependent behaviour, to more detailed mechanisms such as the AIM hypothesis, and finally the behaviour of mirror neurons. I suggested that it is not enough to be content that the behaviour of one's model is compatible with the biological data, but it is necessary to 'push the robotic system' into situations for which we have no biological data. These explicit attempts to generate testable predictions from robotic experiments are essential in order to achieve a fruitful bidirectional flow of information between robotics and the life sciences.

Glossary

Internal model A computational structure simulating an aspect of the modelled system; internal forward models for example simulate the kinematics and/or dynamics of the modelled system, and predict the sensory feedback of an executed action.

Matched dependent behaviour A social learning behaviour where rats negotiating a maze were receiving reinforcement if they matched the actions of a leader rat (Miller and Dollard, 1941).

Real-physics based simulation An approach to simulation that incorporates into the model elements of the physics of the real world; for example, gravity and friction are taken into consideration instead of relying on idealized simulation environments.

Box 7.1 Modelling using robots: from theory to practice

In the main part of this chapter, I described my experiments using simulated and real robots for modelling imitation mechanisms. In this section, I would like to give an insight into the process of turning theory into practice by describing the steps involved in using robots for modelling. I will group them in two stages: robot platform selection and control software selection.

The mechanism to be modelled provides the biggest constraints for the first stage, robot platform selection. There might be specific morphology and articulation requirements for the robot. For example a bipedal structure with a large number of degrees of freedom would usually be involved in an experiment investigating the development of infant walking, but not for experiments in spatial cognition, for which a simple two-wheel mobile robot with a camera and simple sensors would be sufficient. There is a large number of commercial mobile robots readily available (for example, products from ActivMedia Robotics, K-Team, Sony, among others). More complex morphologies can be accomodated using configurable platforms, including commercial robot kits such as the Lego Mindstorms and Robix Rascal. Inevitably, very complex morphologies will not be accomodated by existing commercial offerings, leaving the experimenter with two choices, either building their own robot, or use simulation packages. The latter packages are available with varying degrees of physical realism. One group of simulators utilizes realistic rigid body simulation SDKs (software development kits). This adds more credibility to experiments involving motor control. These include Dynamechs, ODE,

Box 7.1 *(continued)*

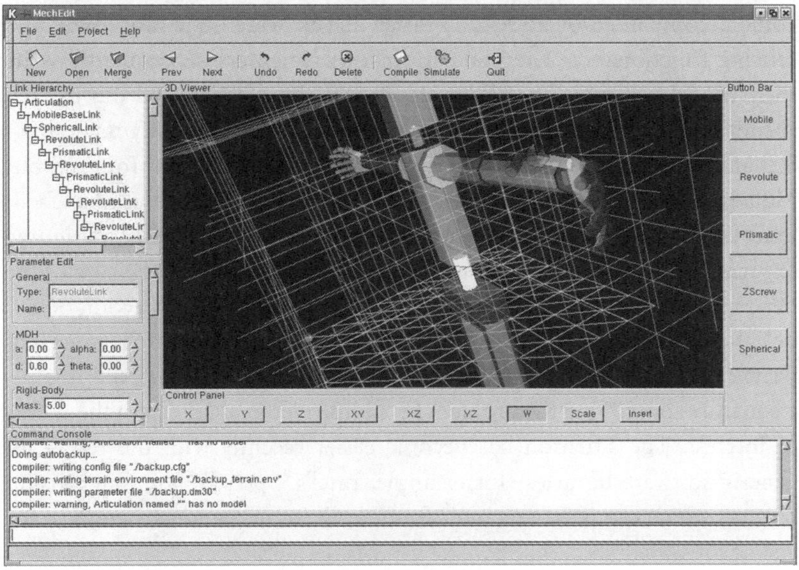

Fig. B7.1 Computer-aided robot design and simulation packages, such as the MechEdit (pictured) can assist in the development of complex robots for experiments in motor control.

and Darwin2k, among others. Source code for these can be found on Sourceforge (http://www.sourceforge.net). There are front-end graphical user interfaces (GUI) for easier construction and simulation of complex robotic systems, for example the MechEdit computer-aided robot design package (see the screen snapshot in Figure B7.1, details can be found in Johnson and Demiris, 2003).

In our case, in the first set of our experiments we used a Lego-based mobile robot as a demonstrator. We used a home-grown mobile robot as an imitator, mainly due to its immediate availability (other mobile robots would also have been suitable). All processing for sensing and control was done onboard the robots, although the option for sending the data to an off-board computer is always available. Similarly, we used a stereo-vision active robot head that was developed at the Electrotechnical laboratory in Japan for the head movement imitation experiments. For the experiments modelling mirror neurons we utilized the Dynamechs real-physics simulation library to construct a complex (13° of freedom) humanoid, since the construction of a real humanoid robot exceeded our capabilities and resources at the time.

For the second stage, the control software selection, the situation is very dependent on the hardware chosen, as most commercial robots will

Box 7.1 *(continued)*

come accompanied by a set of libraries that provide basic movement and sensing functionality. The majority of experimenters subsequently write their model utilizing those libraries (as was the case with all our experiments as well). These are not the only options, and frequently mathematical and/or connectionist modelling libraries and simulators (for example, Matlab, Simulink, PDP++, SNNS, among many others) are used to facilitate the construction of the model. A major difficulty when designing experiments incorporating human–robot interaction (as in the case of the head movement imitation experiments) is the computer vision component; i.e., the processing of images from the cameras. Most researchers tend to write customized software to suit the particular needs of their experiments, frequently simplifying the problem by adding colour markers on the points of interest. The situation has become easier recently with the release of generic software libraries (for example, Intel's OpenCV computer vision library) which can provide a basic framework for faster construction of the vision software.

References

Breazeal, C and Scassellati, B (2002). Robots that imitate humans. *Trends in Cognitive Sciences*, 6(11), 481–487.

Carey, DP, Perrett, D, and Oram, MW (1997). Recognizing, understanding and reproducing action. In F Boller and J Graftman (eds) *Handbook of neuropsychology, Vol. 11*, pp. 111–129. Elsevier Science B. V., Amsterdam, The Netherlands.

Dautenhahn, K and Nehaniv, C (2002). *Imitation in animals and artifacts.* Cambridge, MA: MIT Press.

Decety, J (1996). Do imagined and executed actions share the same neural substrate? *Cognitive Brain Research*, 3, 87–93.

Decety, J, Grezes, J, Costes, N, Perani, D, Jeannerod, M, Procyk, E, Grassi, F, and Fazio, F (1997). Brain activity during observation of actions: Influence of action content and subject's strategy. *Brain*, 120, 1763–1777.

Demiris, J (1999). *Movement imitation mechanisms in robots and humans.* Ph.D. thesis, University of Edinburgh, Scotland, UK.

Demiris, J and Hayes, GM (1996). Imitative learning mechanisms in robots and humans. In V Klingspor (ed.) *Proceedings of the fifth European workshop on learning robots*, pp. 9–16. University of Bari, Bari, Italy.

Demiris, J, Rougeaux, S, Hayes, GM, Berthouze, L, and Kuniyoshi, Y (1997). Deferred imitation of human head movements by an active stereo vision head. In *Proceedings of the 6th IEEE international workshop on robot human communication*, pp. 88–93. IEEE Press, New York, USA.

Demiris, Y (2002). Imitation, mirror neurons, and the learning of movement sequences. In *Proceedings of the international conference on neural information processing (ICONIP-2002)*, pp. 111–115. IEEE Press, New York, USA.

Demiris, Y and Hayes, G (2002). Imitation as a dual route process featuring predictive and learning components: a biologically-plausible computational model. In K Dautenhahn and C Nehaniv (eds) *Imitation in animals and artifacts*, pp. 327–361. Cambriege, MA: MIT Press.

Demiris, Y and Johnson, M (2003). Distributed, prediction perception of actions: a biologically inspired architecture for imitation and learning. *Connection Science*, 15(4), 231–243.

Eneje, E and Demiris, Y (2003). Towards robot active intermodal matching. In A Billard and R Siegwart (eds) *Proceedings of the IROS '03 workshop on programming by demonstration*. IEEE Press, New York, USA.

Fadiga, L, Fogassi, L, Pavesi, G, and Rizzolatti, G (1995). Motor facilitation during action observation: a magnetic stimulation study. *Journal of Neurophysiology*, 73(6), 2608–2611.

Flanagan, JR and Johansson, RS (2003). Action plans used in action observation. *Nature*, 424, 769–771.

Gallese, V, Fadiga, L, Fogassi, L, and Rizzolatti, G (1996). Action recognition in the premotor cortex. *Brain*, 119, 593–609.

Hayes, GM and Demiris, J (1994). A robot controller using learning by imitation. In *Proceedings of the second international symposium on intelligent robotic systems*, pp. 198–204. LIFIA-IMAG, Grenoble, France.

Johnson, M and Demiris, Y (2003). An integrated rapid development environment for computer aided robot design and simulation. In *Proceedings of the international conference on mechatronics*, pp. 485–490. John Wiley and Sons, West Sussex, UK.

Karniel, A (2002). Three creatures named forward model. *Neural Networks*, 15, 305–307.

Meltzoff, AN and Decety, J (2003). What imitation tells us about social cognition: a rapprochement between developmental psychology and cognitive neuroscience. *Phil Trans of the Royal Society of London B*, 358, 491–500.

Meltzoff, AN and Moore, MK (1977). Imitation of facial and manual gestures by human neonates. *Science*, 198, 75–78.

Meltzoff, AN and Moore, MK (1997). Explaining facial imitation: a theoretical model. *Early Development and Parenting*, 6(2), 157.1–14.

Meltzoff, AN and Prinz, W (2002). *The imitative mind: Development, evolution and brain bases*. Cambridge University Press, Cambridge, UK.

Miall, RC and Wolpert, DM (1996). Forward models for physiological motor control. *Neural Networks*, 9(8), 1265–1279.

Miall, RC, Weir, DJ, Wolpert, DM, and Stein, JF (1993). Is the cerebellum a smith predictor? *Journal of Motor Behavior*, 25(3), 203–216.

Miller, NE and Dollard, J (1941). *Social learning and imitation*. New Haven, CT: Yale University Press.

Narendra, KS and Balakrishnan, J (1997). Adaptive control using multiple models. *IEEE Transactions on Automatic Control*, 42(2), 171–187.

Perrett, DI, Oram, MW, Harries, MH, Bevan, R, Hietanen, JK, Benson, PJ, and Thomas, S (1991). Viewer-centred and object-centred coding of heads in the macaque temporal cortex. *Experimental Brain Research*, 86, 159–173.

Rizzolatti, G, Fadiga, L, Gallese, V, and Fogassi, L (1996). Premotor cortex and the recognition of motor actions. *Cognitive Brain Research*, 3, 131–141.

Rizzolatti, G, Luppino, G, and Matelli, M (1998). The organization of the cortical motor system: new concepts. *Electroencephalography and Clinical Neurophysiology*, 106, 283–296.

Schaal, S (1999). Is imitation learning the way to humanoid robots? *Trends in Cognitive Sciences*, 3(6), 233–242.

Scott, SH and Kalaska, JF (1995). Changes in motor cortex activity during reaching movements with similar hand paths but different arm postures. *Journal of Neurophysiology*, 73, 2563–2567.

Tillery, SH, Soechting, J, and Ebner, TJ (1996). Somatosensory cortical activity in relation to arm posture: Nonuniform spatial tuning. *Journal of Neurophysiology*, 76(4), 2423–2438.

Vogt, S (1995). On relations between perceiving, imagining and performing in the learning of cyclical movement sequences. *British Journal of Psychology*, 86, 191–216.

Webb, B (2001). Can robots make good models of biological behaviour? *Behavioural and Brain Sciences*, 24(6), 1033–1050.

Wolpert, DM and Kawato, M (1998). Multiple paired forward and inverse models for motor control. *Neural Networks*, 11, 1317–1329.

Wolpert, DM, Doya, K, and Kawato, M (2003). A unifying computational framework for motor control and social interaction. *Phil Trans of the Royal Society of London B*, 358, 593–602.

Zimmer, HD and Engelkamp, J (1996). Routes to actions and their efficacy for remembering. *Memory*, 4(1), 59–78.

Chapter 8

What neuro-robotic models can teach us about neural and cognitive development

Olaf Sporns

Introduction

The brain is a complex network continually engaged in neural dynamics and in interactions with the environment (Sporns *et al.*, 2004). Brain activity is associated with a distinct repertoire of cognitive states, many of which are accessible to consciousness. What renders a complex network of a few hundred billion nerve cells capable of generating cognitive capacities? How does cognition grow as the network builds its connectivity and begins to interact with the surrounding world? Formulating these questions highlights two major theoretical aspects of cognitive development that are fundamental to the work discussed in this chapter: (a) cognition is a 'network phenomenon', a manifestation of neural activity and connectivity; (b) cognition is embodied, a consequence of the dynamic coupling between brain, body and world.

Brain networks undergo extensive growth, refinement and remodelling in the course of embryonic and postnatal development. They remain malleable and plastic for virtually the entire life of the organism. Perceptual, cognitive and behavioural capabilities at any time within an organism's lifeline are tied to the structure of its brain and body. Importantly, the architecture of the brain, as well as the morphology of the body and the statistics of the environment, is not fixed. Rather, brain connectivity is subject to a broad spectrum of input-, experience- and activity-dependent processes which shape and structure its patterning and strengths (Johnson, 2001). These changes, in turn, result in altered interactions with the environment, exerting causal influences on what is experienced and sensed in the future. Thus, whatever effects cognitive states may have is intricately linked to network structure, input statistics and embodiment.

Recent years have seen an explosion of progress in experimental and computational neuroscience, resulting in a huge increase in the amount of knowledge about functional processes in the developing and adult brain. Some of this knowledge is beginning to be integrated with developmental psychology and cognitive science, leading towards the formulation of models and theories of autonomous mental development (Weng *et al.*, 2001), computational developmental psychology (Shultz, 2003) and a resurgence of cognitive and neural constructivism (Quartz, 1993, Mareschal and Shultz, 1996; Karmiloff-Smith, 1998; Johnson and Mareschal, 2001). Many of the contributions in this volume highlight the various research approaches currently being undertaken in these new areas. What many of these approaches have in common is the central role given to development in the emergence of cognition, and the rejection of nativist accounts of the human mind. This theoretical stance has important consequences for strategies in the design of artificial embodied systems capable of intelligent and autonomous behaviour.

This chapter is about some of the past, present and future contributions of neuro-robotic models to our understanding of neural and cognitive development. Most of the chapter is devoted to a discussion of two emerging principles that may be useful components of theories of embodied development. After introducing some of the design features of neuro-robotic models, we first discuss the principle of value, which provides a theoretical and mechanistic framework for learning and development in embodied nervous systems. Value refers to the outcomes or consequences of behavioural interactions between the organism and robot and its environment, which may lead to the encounter of rewarding or noxious stimuli. The neural basis for value signals in the brain is provided by neuromodulation, the activation of the brain's diffuse ascending systems (such as the dopamine system related to the processing of reward) and their numerous synaptic, neuronal and behavioural effects. After discussing the principle of value, we briefly touch upon the issue of information and its crucial role in shaping neural connectivity and development. Throughout the chapter, we will focus on robot studies that help to clarify the notion that embodiment is a crucial ingredient in assessing the value of neural and behavioural patterns as well as in structuring, sampling and processing information.

Design of neuro-robotic models

Coupling brain, body and world

The design of neuro-robotic systems offers a new research methodology to study the relationship between neuronal processes and behaviour (Beer *et al.*,

1998; Sporns, 2002). Such models are ideally suited to capture key aspects of embodiment because their implementation quite naturally involves continuous coupling of neural states and behaviour in real time. Many neuro-robotic systems are physically instantiated, but the development of powerful software packages has allowed the implementation of some sophisticated embodied models that exist only in simulation. There is some disagreement among researchers about whether embodiment requires the physical instantiation of the model in a robot (for a discussion of this issue see Ziemke, 2004). While the issue is controversial, it seems that physical instantiation is not an absolute requirement, as long as simulated models incorporate some aspects of body morphology, movement and external environment. In a developmental context, simulated embodied systems may in some cases be easier to study as larger regions of parameter space can be explored efficiently. Also, some paradigms for self-organization of connection weights such as evolutionary algorithms can more easily be instantiated in a simulated environment. However, physically instantiated embodied models may also have distinct advantages. It is quite difficult (if not impossible) to capture all aspects of a physical environment, or of a robot body, in a simulation. Thus, even sophisticated simulations of embodied agents run the risk of missing unexpected or unanticipated interactions between brain, body and world.

What many (if not most) neuro-robotic systems have in common is that they consist of three integrated components (Figure 8.1; after Beer, 2000): (a) a simulated neuronal model incorporating anatomical and physiological properties of nervous systems, (b) an autonomous robot with a defined body

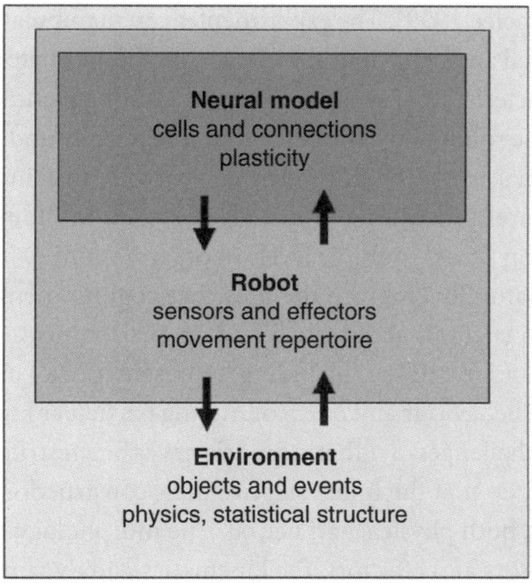

Fig. 8.1 The embodied approach. Neural model, robot and environment are dynamically coupled and continually interact. After a similar diagram by Beer (2000).

morphology and movement repertoire, and (c) the environment containing objects and events. These three components are dynamically and reciprocally coupled. Obviously, neural signals can cause movements of the body and thus action in the environment. It is perhaps less obvious, but equally important to note that the effects of neural states on the environment can have an impact on the statistics and nature of sensory inputs reaching the nervous system. In other words, an embodied system determines what its future inputs will be and thus imposes structure on its own sensory input space (Nolfi and Parisi, 1993; Pfeifer and Scheier, 1999; Schlesinger, 2002). To give a concrete example, consider a robot which can push (physically displace) objects in a visual scene. The action of the robot causes the sudden appearance of spatially and temporally correlated movement in the visual array, which can be used to segment the object from a background (e.g. Fitzpatrick *et al.*, 2003; Fitzpatrick and Metta, 2003). Another sensorimotor function with very dramatic effects on the statistics of visual inputs is attention-driven saccades (camera/eye movements) which determine the direction of gaze (e.g. Breazeal *et al.*, 2001). Depending on the state of the attentional system different kinds of objects are preferentially selected and placed in the central region of the visual array where they may be subjected to closer visual analysis. Action and perception of neuro-robotic systems form closely coupled dynamical loops and are often inseparably linked. We will return to this crucial aspect of embodied systems later in this chapter.

Synthetic modelling

The design and construction of an artificial embodied system often provides unique insights into the relationship between brain and behaviour (Reeke and Sporns, 1993; Pfeifer and Scheier, 1999). The experimenter can manipulate and record a broad range of neural and behavioural variables, including anatomical pathways, biophysical characteristics of synapses, neuronal tuning parameters, physical characteristics of the robot body, the behavioural repertoire and stimuli and events in the external world. Such models can provide insights into normal brain and behavioural processes and can also be used to illuminate brain disease and dysfunction across multiple levels of organization.

Synthetic models quite naturally embrace the idea that cognition emerges from interactions between a (neural) architecture and an environment. More complex neural properties and functions (including 'representations') appear as a result of experience and behaviour and of encountering particular kinds of sensory stimuli and motor challenges. Synthetic models never begin structureless, as a tabula rasa. In fact, even at the initial stage of their construction they contain abundant structure, both physical and neural. The morphology of the body, the capabilities of sensors and effectors, the kinematics and dynamics of the motor repertoire all impose severe constraints on the kinds of interaction between the system and the environment that are possible. The structure of the nervous system, the topology of its connection pattern, the biophysical properties of synapses and cells, the capacity of connections to change or remain constant all define a specific set of structural parameters. The structure of any embodied system strongly deviates from randomness, and thus, in a statistical sense, incorporates a huge amount of information right from the start.

While synthetic models contain lots of structure, this does not amount to an endorsement of the nativist paradigm. Rather, it is absolutely essential that their structure be modifiable and expandable to deal with aspects and challenges of the environment that cannot be 'wired in' from the beginning. In this chapter, we will focus entirely on modifiable neural structures, and leave aside the interesting domain of modifiable body structure (Hara and Pfeifer, 2003). Neural structures may be modified through activity- or experience-dependent processes, which affect the strengths of existing connections or control their growth and structural elaboration. This modification of neural structures, viewed at a large scale and over longer developmental periods, is not a random process, but serves the overall goal of increased performance of the organism or robot. In order to ultimately generate learning and increased adaptation, there needs to be 'matching' between the (physical and neural) structure of the synthetic model and the structure of the environment. This matching requires that internal and external structure be compared and continuously updated. Viewed from the perspective of matching, the dialogue between the embodied system and the environment becomes the motor of development and defines its global trajectory.

Learning and development

Can we come up with a quantitative, mechanistic framework for what learning and development accomplish in terms of neural information? Neural structures can be conceptualized as networks that generate and integrate information (Tononi *et al.*, 1998; Tononi and Sporns, 2003). Neural activations express and capture statistical features of their inputs (arriving from the environment via sensory channels, or from other neurons via connection pathways), thus generating information. This information must not remain isolated but needs to be integrated to allow the emergence of coherent cognitive states and behaviour utilizing multiple distributed sources of information.

Learning and development operate on the two main dimensions of neural function, neuronal activity and connectivity, to alter the informational capacity of the brain in accordance with the particular statistics of the environment that this brain is exposed to. There is a plethora of candidate neural mechanisms, ranging from changes in the numbers of cells and synapses, to alterations in cellular neuronal morphology and synaptic plasticity (Singer, 1995; Katz and Shatz, 1996; Sur and Leamey, 2001). Cortical growth and maturation, including myelination of long-range connections (e.g. Paus *et al.*, 1999), may also contribute to the capacity of neural tissue to acquire new response properties and integrative capacities. A variety of biological mechanisms of learning and development have been analysed in computational models, ranging from traditional connectionist models, where learning is often implemented as a process of convergence to a specific end state, to computational neuroscience models that are validated by direct comparison with results from experimental neurobiology.

Structural and activity-dependent changes in the nervous system do not occur indiscriminately but rather follow very specific spatial and temporal patterns. The timing of stimuli and their statistics are important ingredients in 'directing' learning-related changes such that improved performance in a specific task becomes integrated within the overall behavioural repertoire of the organism. Value and information are important candidate principles for learning and development which we will now explore.

Value: neuromodulation and learning

Value

Why is value such a central principle in development? Reeke *et al.* (1990) conceptualized value as imposing biases 'on the outcome of previous interactions with the environment'. In order for a system to be adaptive, value signals are needed that 'reflect the global evaluation of recent behaviour'. Friston *et al.* (1994) stated that 'the value of a global pattern of neuronal responses to a particular environmental situation (stimulus) is reflected in the capacity of that response pattern to increase the likelihood that it will recur in the same context'. This definition of value highlighted its conceptual similarity to the concept of fitness in evolutionary adaptation. If we consider that neuronal response patterns can be mapped onto behaviours, value then defines the shape of an adaptive landscape (Sporns and Edelman, 1993) with peaks of high value (associated with 'valuable' neuronal response patterns and their concomitant behaviours) and valleys of low value (associated with neuronal patterns that do not generate value). The shape of this value landscape is determined by external factors related to the environment within which behaviour occurs and to the body structure of the behaving system, as well as by internal needs or 'biases'.

Value has a number of properties that set it apart from other learning mechanisms such as Hebbian learning, or back-propagation. It is essential that value acts as (a) a *global signal* which is (b) *internally derived* by the behaving system, (c) *after actual behaviour* has occurred. Thus, value is essentially embodied, tied to the physical structure of the robot or organism and to its actions as they unfold in time. Unlike the 'error signals' central to some learning algorithms such as back-propagation, value is determined in an unsupervised (or self-supervised) manner. The value signal reflects global performance of the organism, not the differential contributions of cells or synapses relative to a 'desired output'. Value is most closely related to reinforcement learning, with a few notable differences. Unlike in most formulations of reinforcement learning, value is not derived or picked up from the environment, but is determined within the behaving system itself as a function of current and previous sensory inputs, as part of the neural and bodily context. As discussed in Friston *et al.* (1994), value-dependent learning is formally related to a variant of reinforcement learning, called temporal difference learning.

Many studies suggest that some environmental stimuli and events, such as rewards, noxious or painful stimuli, or stimuli that are novel or violate expectations have a greater impact on learning and development. Such highly salient inputs may include touching an object, high-contrast visual features such as blobs or edges, or moving and colourful stimuli (reviewed in Thelen and Smith, 1994). In human infants, sensitivity to highly salient sensory inputs emerges very early in postnatal development and seems capable of driving at least some forms of learning and development. The incidence of behaviours that lead to the occurrence of such salient stimuli tends to grow (e.g. Angulo-Kinzler, 2001), as these behaviours are highly 'valuable'. Value can thus act to change the behavioural repertoire of an individual organism in a manner that channels behaviour towards regions of the adaptive landscape that produce more salient consequences.

Value systems

Value systems serve as internal correlates and mediators of value and saliency and consist of well-defined neural substrates that are found in virtually all vertebrate species. The functional roles of value systems have been studied in a number of computer simulations and robot models. Value systems were integral components of the Darwin series of developing autonomous robots (Reeke *et al.*, 1990; Edelman *et al.*, 1992; Almassy *et al.*, 1998; Sporns *et al.*, 2000). The Darwin robots learned to perform simple sensorimotor tasks (foveation and reaching) as well as visual and visuotactile categorization. Value systems have also been employed in the SAIL robot (Huang and Weng,

2002) modelling habituation, reinforcement and novelty. Several robot designs incorporate motivational systems whose functions are related to (although generally more complex than) those of value systems. The control architecture of the humanoid robot Cog contained a motivational system regulating its behaviour, particularly its social interactions with humans (Adams *et al.*, 2000). Similarly, drives and emotions were related to the needs and states of arousal of Kismet (Breazeal, 1998, 2002), a robot capable of social interactions with humans. The motivational system was intricately coupled to the robot's attention module and to processes involved in behavioural selection and control, especially those related to human–robot interactions.

Value systems generate signals that are used to adjust the probabilities of behaviours by modulating synaptic changes within neuronal networks through value-dependent learning. As mentioned above, a key difference with reinforcement learning is the explicit connection of value systems with specific neural structures and processes. Value systems have several characteristic anatomical and physiological properties: (a) they act by modulating neural activity or synaptic plasticity; (b) they exert their effects by delivering a diffuse, global signal over extended brain regions. In typical implementations, value systems are incorporated in the neural network architecture of the agent (robot or organism) and become active after the occurrence of specific sensory stimuli, which are often encountered as a consequence of behavioural actions. The activation of these value systems is short-lasting (phasic) and constitutes a timing (gating) signal for synaptic modification.

What constitutes a valuable or salient stimulus is itself subject to change over the course of learning and development. Value systems cannot rely on prewired inputs from sensory regions to generate their signals, but must incorporate activity- and experience-dependent processes. Friston *et al.* (1994) introduced the distinction between innate and acquired value (see the schematic in Figure 8.2). Innate value was viewed as evolutionarily determined, similar to an innate bias. Behaviours that satisfy homeostatic or appetitive needs, consummatory activities, or avoidance of noxious stimuli predominantly reflect priour evolutionary selection and are therefore in most cases independent of learning or experience. Such innate value, however, cannot reflect the specific configuration of the environment and cannot include stimuli that are themselves initially neutral but, in a specific environmental context, become predictive of future valuable events. Acquired value is activity-dependent and allows the value system to become sensitive to stimuli that are not by themselves able to trigger a value-related response.

Modulatory neurotransmitter systems, including dopamine, serotonin and acetylcholine, possess many of the structural and functional properties of value

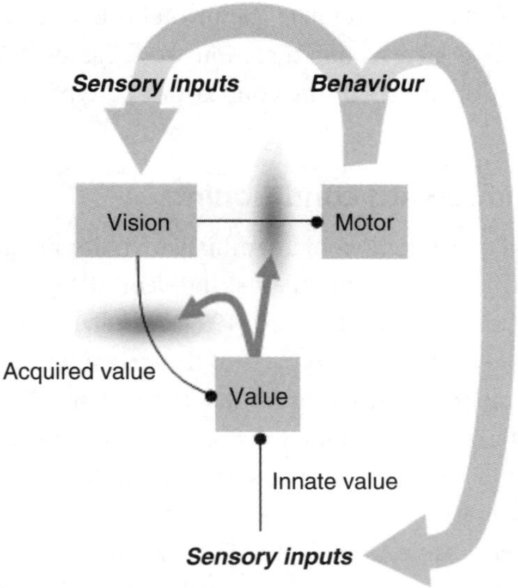

Fig. 8.2 Schematic diagram illustrating the action of value systems. Neural signals in 'vision' trigger 'motor' activity and behaviour. Behaviour in turn results in sensory inputs. Some sensory inputs are relayed through sensory afferents to the value system where they elicit a neuromodulatory (value) response. The value signal modulates visuomotor connections, as well as sensory afferent from 'vision' to 'value'. Future occurrences of the visual patterns preceding the encounter of the innately valuable sensory inputs becomes capable of triggering a value response (acquired value). Grey boxes = neural networks for vision, motor and value; thin lines = excitatory connection pathways; shaded arrow = neuromodulatory (value) connections; shaded ellipses = projection targets of the value signal (local neuromodulator concentration) where value-dependent learning takes place. Note the distinction between sensory inputs to the value system that mediate innate and acquired value.

systems, including their capacity to acquire new response characteristics in the course of experience. In biological nervous systems, neuromodulators have an extremely broad range of functions including the regulation of neuronal excitability, effects on the expression of cellular proteins, structural modifications in neurons and neural circuits, as well as modulation of synaptic plasticity (Hasselmo *et al.*, 2002). Several of these functions have been implemented in neurocomputational models, primarily as effects on neuronal response functions, learning rates, or other model parameters (Hasselmo, 1995; Fellous and Linster, 1998). From the perspective of neural and cognitive development, the potential functional roles of diffusely projecting neuromodulatory systems in influencing the magnitude and direction of synaptic plasticity are of particular importance. Synaptic plasticity is a main mechanism for generating lasting behavioural and representational change. A developing robot, built on biological principles, should emulate the pivotal role of neuromodulation in real brains. As a first step in this direction we focus on the actions of one neuromodulator (dopamine) in the context of one type of learning (reward conditioning).

Dopamine and reward conditioning

Recent studies of the function of mammalian midbrain dopamine system implicate one of its major components, the ventral tegmental area (VTA), in reward conditioning (Schultz *et al.*, 1997; Schultz, 1998). Dopaminergic neurons within the VTA project to parts of the striatum, the amygdala and to widespread cortical areas, including frontal and prefrontal cortex. Dopamine modulates long-term potentiation, a major candidate mechanism for synaptic plasticity in the central nervous system. Long-term potentiation (LTP) and long-term depression (LTD) have been shown to occur in some of the afferent and efferent connections of the VTA, including plasticity of excitatory inputs to both dopaminergic (LTD) and GABAergic (LTP) cells of the ventral tegmental area (Bonci and Malenka, 1999). Also, dopamine modulates synaptic plasticity of excitatory synapses in both the VTA and the nucleus accumbens (Thomas *et al.*, 2000; Overton *et al.*, 1999). The functional significance of this type of plasticity in the context of reward-related behaviours is not yet fully understood. In addition to their potential roles in reward conditioning, dopamine projections to cortical areas have a significant impact on the functional topography and representations of cortical maps (Bao *et al.*, 2001).

VTA firing patterns undergo highly characteristic changes in the course of reward conditioning. Prior to conditioning, VTA dopamine neurons show a phasic response to food rewards. Immediately after the food reward is received, dopamine neurons discharge a brief burst of spikes lasting no more than a few hundred milliseconds. In the course of learning, this response pattern changes in a characteristic manner. When a reward is preceded by stimuli which reliably predict the occurrence of the reward, dopamine neurons no longer respond to the occurrence of the (already predicted) reward. Rather, their phasic activation is 'transferred' to the onset of the reward-predicting stimulus, while activation at the occurrence of the primary (now completely predicted) reward becomes attenuated or disappears entirely. Moreover, if a reward is fully predicted but does not actually occur, there is a transient depression in the baseline firing rate of dopamine neurons at the time of the expected occurrence of the reward. Several models of the midbrain dopamine system have been proposed (Montague *et al.*, 1996; Schultz *et al.*, 1997; Suri and Schultz, 2001; Suri, 2002), emphasizing the relationship between dopaminergic responses and temporal difference learning.

While neuromodulatory systems related to reward, such as dopamine, normally mediate the effects of naturally occurring rewarding stimuli, their function is altered in a number of brain diseases, including drug addiction (Wise, 1996; Berke and Hyman, 2000). Addiction is a chronic disorder, characterized by a persistent state of compulsive drug use, intense drug craving and cue-initiated relapse, often accompanied by the experience of serious negative consequences. Which specific brain processes cause the transition from normal reward behaviour to full-blown drug addiction is still largely unknown. A neurobiological hypothesis states that drugs may cause reorganization of neural circuitry, either through effects on synaptic strength, neuronal excitability or neuronal morphology (Hyman and Malenka, 2001). Supportive evidence suggests that cocaine exposure can cause rapid and marked increases in synaptic strengths in midbrain neurons (Ungless *et al.*, 2001), and that drug-related plasticity can contribute to relapse in drug-seeking behaviour (Vorel *et al.*, 2001). One of our goals in exploring reward processing in neuro-robotic models is to develop new hypotheses regarding the origin of addiction.

Structure of a neuro-robotic model of neuromodulation in learning

Because of the central role played by dopamine in a variety of cognitive processes, numerous computational models of dopamine function have been proposed. A comprehensive, or even casual, review of these models is beyond the scope of this chapter. Here we present a brief summary of some of the results we obtained by implementing a dopamine neuromodulatory system in a neuro-robotic model. In discussing our results we will focus on the issue of embodiment and the role it plays in the development of the model. In this chapter, we only provide a brief description of the design of the robot, its environment and the neural model (see Box 8.1), noting that they are described in detail in several previous publications (Sporns and Alexander, 2002; Alexander and Sporns, 2003). The interested reader should consult these original publications for additional detail.

Neuromodulation in an autonomous robot

In testing our model we used computer simulation in addition to implementing the model in an autonomous robot. First, we presented stimuli in relatively fixed temporal patterns, reproducing conditions in animal experiments in which stimuli and rewards are separated by a constant time interval. In robot

Box 8.1 A robot model of neuromodulation

General

We used a mobile robotic platform (a Khepera robot; Figure 8.3), interfaced with a neural simulation implemented in Matlab and running on a fast workstation. In order to be able to control sensory stimuli, the robot is located in an environmental enclosure. For experiments on reward conditioning, the environment contains coloured (red) objects, all of which have a non-conductive surface. All neural and synaptic states are recorded for each experiment and analysed and displayed offline. In addition, an overhead camera records views of the environment and of the robot's actions at a rate of approximately 1 frame/second.

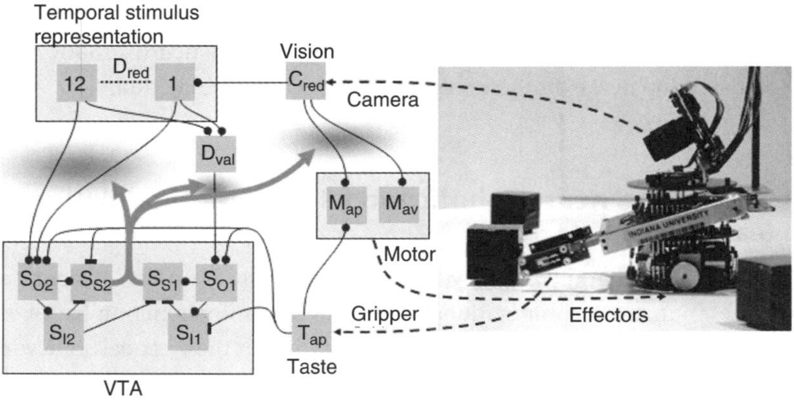

Fig. 8.3 Robot (right) and schematic diagram of neural model (left). (Modified after Alexander and Sporns, 2003.) Neural network components are indicated as shaded boxes, thin lines = excitatory connections; thick lines = inhibitory connections; shaded arrow = value signal (V_R); shaded ellipses = projection targets of the value signal. Components are vision, the temporal stimulus representation, motor, taste and VTA. The VTA consists or several excitatory and inhibitory units which generate positive and negative prediction errors that are then combined into V_R. Connections between units in D_{red} are omitted in this figure for clarity; there are a total of 12 units in D_{red} with different temporal receptive fields.

Robot

The robot ('Monad') is capable of navigating through the environment, avoiding obstacles (walls) guided by IR sensors, and of physically contacting objects using a moveable arm and gripper. Main sensors interfaced with the neural model are vision (through a colour CCD camera mounted on

Box 8.1 *(continued)*

the robotic platform) and taste (conductivity of objects). Low conductivity activated appetitive taste receptors, corresponding to the occurrence of a reward.

Environment

The enclosure was about 1 metre on each side and contained red (appetitive) objects. These objects were placed in the environment by the experimenter at the beginning of each run. The objects were light and could easily be moved and pushed around by Monad in the course of an experiment. One of the environmental parameters that could be manipulated by an experimenter was the number (density) of objects. We conducted experiments with low object density (four objects total) and high object density (12 objects total).

Behaviour

Visual detection of objects (described in detail in Sporns and Alexander, 2002) resulted in approach behaviour under visual guidance. Once an object was in close proximity, the arm was lowered and the gripper closed, while Monad turned slightly to align the gripper surfaces with the object. Sensing of the 'taste' of the object was followed by its subsequent release and the continuation of environmental exploration. The encounter of reward in conjunction with the gripping of red objects strengthened connections between appropriate visual and motor units, rendering vision capable of triggering the behavioural sequence leading to physical contact with ('consumption of') red objects. Since only rewarding objects were used in the present set of experiments, there is little or no overt change in behaviour (i.e. the unconditioned response is very similar to the conditioned response after learning). Note that other experiments (Sporns and Alexander, 2002) using aversive objects yielded different behavioural patterns (approach and avoidance) in the course of learning.

Neural model

All neural units were implemented using a continuous firing rate model with a single saturating non-linearity, according to

$$s_i(t+1) = \phi[A(t) + \Omega s_i(t)] \tag{1}$$

where $s_i(t)$ is the activity of unit i at time t, $A(t)$ is the total synaptic input to unit i at time t, Ω is the unit's temporal persistence ($0< \Omega <1$), and $\phi[.]$ is the saturating nonlinearity (both hyperbolic tangent and sigmoidal

Box 8.1 *(continued)*

functions are used). $A(t)$ was calculated as the linear sum of all inhibitory and excitatory inputs, i.e. $\Sigma c_{ij}s_j(t)$. The neural model of the mammalian midbrain dopamine system was aiming to incorporate several functional and anatomical characteristics of the real neurobiological structures (Figure 8.3). In the neural model, several types of cells comprised an area corresponding to the ventral tegmental area (VTA), whose outputs modulated plasticity in several different locations. Input to the modelled VTA was provided by a temporal representation of the visual stimulus (representing a highly simplified analogue to prefrontal cortex, PFC). Colour-selective units (C_{red}) relayed sensory inputs to a network transforming this input into a continuous temporal representation. As a result of excitatory and inhibitory interactions within this network (essentially forming a delay chain), stimulus-specific units D_{red} became active after a specific amount of time (between 1 and 12 iterations) had elapsed since the onset of their preferred stimulus. These units had fairly broad temporal tuning with significant mutual overlap in terms of their 'temporal receptive field'. D_{red} units projected to the VTA (for more detail on the structure of the VTA and its connectivity, see Sporns and Alexander, 2002; Alexander and Sporns, 2003), making a series of 12 modifiable connections. The overall level of neuromodulator released by the VTA reward system was taken to be a 'value signal' or reward signal V_R. V_R was a signed scalar variable, which was positive if the level of neuromodulator increased above a baseline and negative if it decreased below a baseline. The neural model used this value signal in value-dependent learning to adjust all connection strengths. Through value-dependent learning, the value signal V_R influenced synaptic modification in three sets of connections. Of particular importance in this paper are the 12 connections linking the temporal delay units D_{red} to the VTA. Through value-dependent modification of these connections, value (reward) was able to modify the response characteristics of parts of the neuromodulatory system (the VTA) itself. Connections that were subject to value-dependent learning were updated according to a ternary learning rule:

$$c_{ij}(t+1) = (1-\varepsilon)c_{ij}(t) + \eta s_j(t) F(s_i(t)) V_R \qquad (2)$$

where c_{ij} = connection weight from unit j to unit i, ε = incremental decay rate of connection weight per iteration, η = learning rate, $F(.)$ = nonlinear function applied to postsynaptic activity, V_R = value signal. The value signal represents the local concentration of neuromodulator (here

Box 8.1 *(continued)*

dopamine) released as a result of the activity of the value system. $F(.)$ determined if a connection weight increased or decreased, depending upon the level of postsynaptic activity. $F(.)$ was a continuous saturating function ($-1 < F(.) < 1$) with a temporal profile that reflected the dependence of synaptic modifications on postsynaptic activity.

experiments, constant timing can be achieved by manually presenting objects, or by conducting experiments in environments with low object density. When using such discrete learning trials, both computer simulation and experiments conducted with the robot yielded results that were consistent with the observed physiological characteristics of the mammalian dopamine system (Figure 8.4). The model was able to reproduce a broad spectrum of phenomena, including: (a) dopamine responses to unpredicted rewards; (b) transfer of the dopamine response to reward-predicting sensory stimuli (as a result of synaptic plasticity in afferents to the dopamine system itself); (c) inhibition (attenuation) of the dopamine response at the time of omission of a previously predicted rewarding stimulus; (d) dopamine responses to reward-predicting and predicted stimuli that occur either early or late; and (e) extinction. Phenomena (a), (b), (c) and (e) are illustrated in Figure 8.4 (for more data see, e.g. Alexander and Sporns, 2003).

However, little experimental data is available on the performance and temporal characteristics of the dopamine system in the context of unconstrained and autonomous behaviour. In robot experiments with a high density of objects present in the environment, we noticed that the displacement of objects that resulted from the robot's behavioural activity quickly produced non-uniform and highly clustered spatial distributions of reward (Figure 8.5). This spatial distribution altered the statistics of stimulus encounters (Figure 8.6). At the beginning, objects were still well separated and relatively constant temporal relationships between visual inputs and subsequent rewards allowed the emergence of synaptic patterns that were consistent between timed trials (Figure 8.6a, right) and autonomous behaviour (Figure 8.6b, right). As the experiment progressed, however, objects were pushed together and the timing of successive object encounters was more irregular. Specifically, rewarding objects were often encountered in rapid succession, as the robot was 'attracted' to object clusters. Synaptic patterns reflected this altered timing of reward and developed marked differences to patterns obtained with timed trials.

This simple set of experiments provided an indication of the potentially close and reciprocal interactions between neural plasticity and behavioural activity.

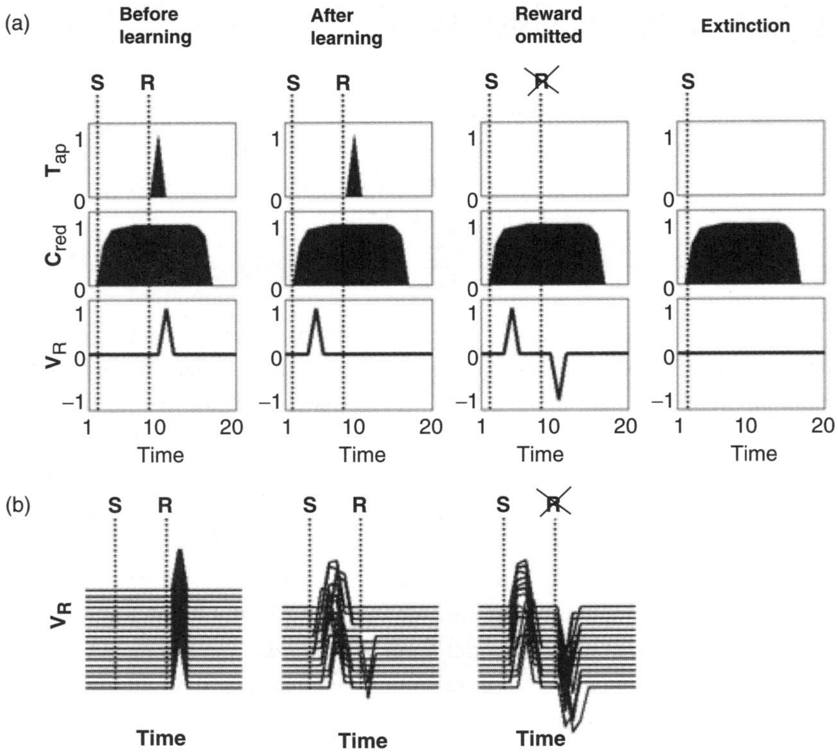

Fig. 8.4 Plasticity of dopamine responses in the simulation (a) and in the neuro-robotic model (b). Activity traces of appetitive taste (T_{ap}), colour-selective visual units (C_{red}), and the value signal V_R are shown. Markers 'R' and 'S' signal the time of onset of the primary reward (appetitive taste) and the visual appearance of the red object, respectively. Before learning, S is neutral and R is unpredicted. R triggers a phasic response of V_R. After S and R have been coupled for several trials, S now predicts the occurrence of R. S elicits a V_R response, R does not. If, after learning, the predicted reward is unexpectedly omitted, S still elicits a V_R response, but at the time of the expected reward, V_R is decreased below zero (baseline). If the reward is omitted for some time, the reward prediction and V_R response is extinguished. The plots in panel B show stacks of traces of V_R (dopamine or value signal) that are temporally aligned to the onset of taste (R) for sets of 15–20 individual object encounters (trials). These response traces were obtained in robot experiments conducted with discrete learning trials, in which a single object was presented manually, allowing consistent timing between visual appearance of the object and reward delivery. Note that, under these conditions, the pattern of dopamine response (corresponding to the value signal V_R) is virtually identical in simulations (a) and in the neuro-robotic model (b).

The differences we observed between timed trials and autonomous behaviour in environments with high reward density were unexpected and might not have been obtained if we had implemented the model only as a simulation with fixed stimulus sequences and timings. There are two main reasons for the

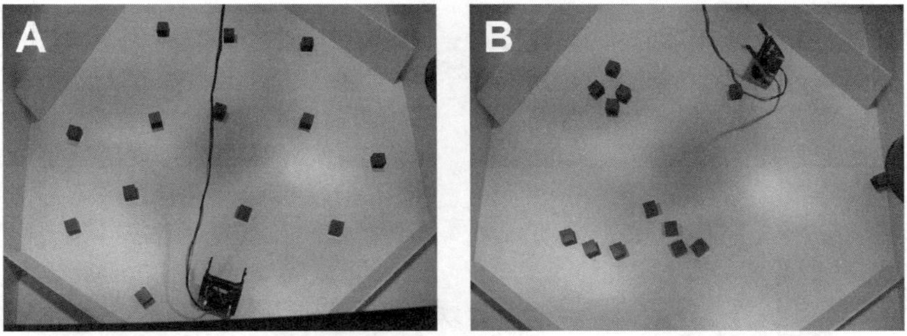

Fig. 8.5 Spatial distribution of objects at the beginning (a) and at the end (b) of a typical robot experiment. Motor actions of the robot resulted in a clustered distribution of reward (objects).

unexpected nature of these findings. First, the robot is navigating from object to object choosing its own path of approach, which introduces a temporal 'jitter' into the relative timing between the onset of the reward-predicting stimulus and the reward itself. Second, the actions of the robot alter the spatial distribution of the rewarding objects by increasing their local density, which in turn further degrades the precision of the temporal prediction. As we will argue now, this type of interaction between behaviour and plasticity form an important rationale for conceptualizing and studying development and learning as an embodied process.

Summary

Our robot experiments demonstrate that neuromodulatory systems are essential in linking behaviour and neuroplasticity. In the model, a 'value signal' was used to modulate the strengths of synaptic connections, thus implementing the observed functional effects of neuromodulators in neural plasticity. Importantly, value acts as a global signal affecting widespread projection areas through a burst of phasic and short-lasting activation. Thus, the signal combined spatial uniformity and temporal specificity. These properties of value systems correspond closely with the anatomical and physiological characteristics of mammalian neuromodulatory systems, including the dopamine system. This correspondence provides a promising rationale for the incorporation of such systems in computational models implemented in autonomous robots. The action of the value system in the present model, through changes in sensorimotor connections and inputs to the neuromodulatory system itself, allows an autonomous agent to learn and adaptively change its behaviour without an external 'teacher'.

Interesting differences emerged in our model when comparing neural development during timed trials (in which behaviour is fairly stereotypic) and

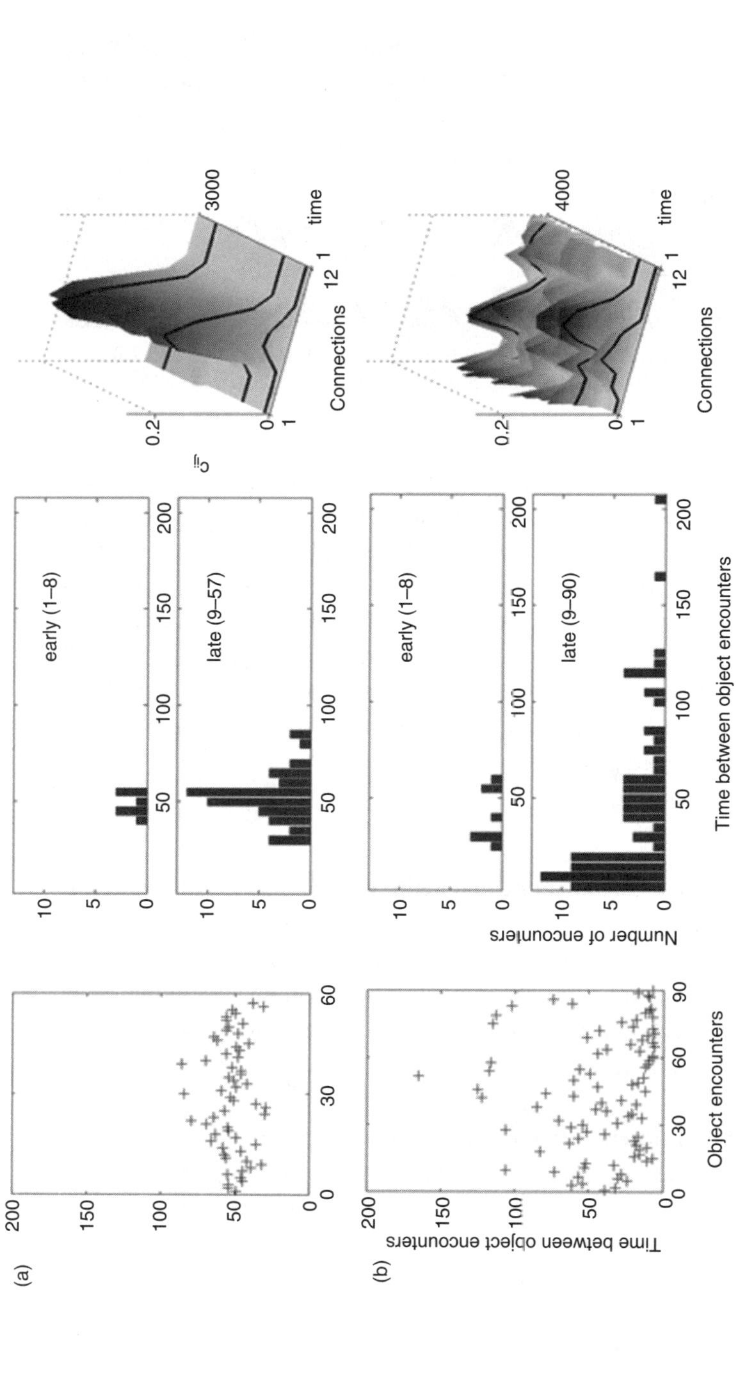

Fig. 8.6 Changes in the statistics of object encounters and synaptic patterns during autonomous behaviour. Panels in (a) show data from a robot experiment with timed learning trials (manual feeding of objects). Panels in (b) show data from a robot experiment allowing autonomous behaviour and involving the spatial redistribution of objects (see Figure 8.5). Plots show time series of object encounters (left) and the distribution of time intervals between successive encounters (middle), as well as the temporal development of synaptic strengths for the 12 connections linking the temporal stimulus representation to the VTA (right). In the plots of synaptic strengths, black lines indicate the connection profiles at 100, 700 and 2500 (3500) time steps.

during autonomous behaviour. In experiments involving autonomous behaviour, we observed a time-dependent alteration of an important environmental variable, the spatial distribution of reward throughout the environment, due to the behavioural activity of the robot. This alteration, in turn, had consequences on synaptic patterns encoding predictions about the occurrence of future rewards. The differences in spatial reward distribution between early and late phases in experiments with high object densities were not the result of purposeful rearrangements of the environment by either robot or experimenter. Instead they were a 'by-product' of sensorimotor activity displayed by the robot, its tendency to push objects around while interacting with them until they form clusters—see also the didabots of Maris and te Boekhorst (1996) and Scheier *et al.* (1998) for similar effects in very different robot experiments. This spatially non-uniform distribution of reward was the outcome of the coupling between brain, body and environment. As this example illustrates, this coupling was strongly reciprocal. Behaviour affected the statistics of reward timings which drove synaptic plasticity through activation of the neuromodulatory value system. In turn, synaptic changes altered the coupling between visual and motor units which affected subsequent behaviour.

Information: shaping input statistics

The discussion and the examples presented above of the action of value and neuromodulation in learning revealed a surprising degree of coupling between neural and behavioural variables. A robot's actions altered the spatial distribution of rewarding objects, which in turn impacted on the timing of rewarding stimulus encounters. This simple example highlights a fundamental developmental issue, to which we now turn briefly. We suggest that robots and organisms do not passively absorb information from their surrounding environment, but rather their actions *on* the environment select and shape the information.

The statistics of sensory inputs are of great importance in learning and development. Early sensory areas of the brain, including visual, auditory and somatosensory cortex, show a high degree of experience-dependent plasticity. Cortical representations reflect the history of sensory stimulation (e.g. Kilgard *et al.*, 2001) as well as motor actions and behaviours (e.g. Nudo *et al.*, 1996). In severe cases of sensory deprivation or over-stimulation the development of normal sensory capacities may be disrupted. The sensitivity of brain tissue to the distribution and temporal sequence of inputs gives special significance to any causal influences of embodiment on input statistics.

Neuro-robotic models have provided several examples of how the development of specific neural and cognitive functions may depend on the actions of a robot. Scheier and Lambrinos (1996) and Pfeifer and Scheier (1999) have developed several models of perceptual categorization that utilized robot behaviour to generate inputs allowing the discrimination of objects belonging to different categories. Almassy *et al.* (1998) modelled the development of complex receptive field properties in visual cortex in a behaving robot. They found that self-generated movements of the robot resulted in smooth lateral displacements of objects within the robot's visual field, thus generating temporal correlations over multiple image frames that could be exploited by neurons in inferiour temporal cortex. In addition, the map of modelled neurons in inferior temporal cortex showed experience-dependent fluctuations which reflected the history and frequency of stimulus encounters. Kritchmar *et al.* (2000) extended this aspect of the model and recorded systematic changes in receptive field properties of object-selective visual neurons with object composition of the environment, demonstrating the experience-dependence of perceptual categorization.

Can we measure the extent to which sensory information is structured and generated? Extensive analyses of static visual data have shown that natural images have highly characteristic statistical structure (Simoncelli and Olshausen, 2001). Some of the methods used to characterize natural images can be applied to input streams obtained from robots or living creatures. For example, appropriate information-theoretical measures could be applied to streams of sensory data gathered using different sensorimotor strategies. Initial analyses have shown that simple sensorimotor functions like gaze direction and foveation can generate high mutual information and complexity in visual inputs (Lungarella and Pfeifer, 2001; Sporns and Pegors, 2003; Lungarella *et al.*, 2005). Such informational patterns can be exploited by neural circuits and promote the stabilization of matching neural connections that incorporate recurrent statistical features. Thus, the structuring for learning and development in robots and organisms of sensory inputs through embodied action may emerge as a key principle (Sporns and Pegors, 2004).

Lessons learned from studying neuro-robotic models

What have we learned from our studies of neuro-robotic models about the specific constraints imposed by brain architecture and function, and by embodiment on neural and cognitive development? Our studies have pointed to the central role of value systems in signalling the consequences of behaviour and influencing the development of synaptic patterns in widespread regions of the brain. Value systems are part of the brain itself. They provide means by which a specialized part of the brain can influence other brain regions to influence learning and development. It was essential in our model that value systems could modify their own inputs and thus show flexible and dynamic response properties. Value systems are also essential components of embodied systems, because they allow the system to autonomously evaluate its own (embodied) actions. Its actions may also modify its value system, through changes in connections that link sensory representations to the value system itself.

Our robot experiments lead to several experimental predictions regarding the activity and functional role of neuromodulatory systems (in particular, dopamine) in the course of 'natural', self-guided behaviour. We found that clusters of rewarding objects exerted an 'attractive force' on the robot, which resulted in restricted trajectories of robot movement and navigation as well as repeated 'rapid-fire' sequences of reward encounters. Disruptions of the neurobiological bases of reward processing are thought to form a major cause for lasting behavioural changes and, eventually, chronic disease (addiction) in humans. Our results suggest the hypothesis that a pattern of persistent reward-seeking behaviour may in part be generated as a result of a progressive reshaping of the environment coupled with long-lasting synaptic changes in specific neural structures. Future robot experiments will investigate this hypothesis in detail.

The effects of behaviour and motor activity on shaping the statistics of sensory inputs provide a fundamental theoretical argument for the importance of *embodied* cognition. The observations reported in this paper depended critically on behaviour carried out in a real environment. They could not have been predicted from the consideration of timed trials alone or from computer simulations that do not explicitly incorporate the coupling between brain and behaviour. Rather, the fact that the neuro-robotic system, through its embodied actions, was altering the stimulus distribution in the environment had a significant impact on the development of synaptic patterns and neural representations.

Conclusion

Can artificial cognition be created without development? Numerous research groups around the world are engaged in the design of robots or sophisticated software agents with the explicit goal of creating systems that are capable of some form of cognition, perhaps approaching or replicating that of humans. This research effort is in the middle of a very significant paradigm shift. Classical AI, for the most part, attempted to create intelligence and higher cognition directly, by crystallizing knowledge and formalizing cognitive principles mostly obtained from the study of adult human beings. Essentially, this approach assumed that it is possible to side-step development entirely, relegating development to the role of a necessary but non-essential transient. An increasing number of researchers in AI, robotics and autonomous systems are diverging from this non-developmental approach. Instead, development is becoming the central research issue in the ongoing endeavour of creating machine intelligence. Weng *et al.* (2001) envision that the creation of robots with human capabilities will depend on developmental processes and 'programs' and that sophisticated cognitive functions will emerge from prolonged interactions between robots and their environments, including human teachers.

Addressing the challenges of development will require new constructive, integrative and synthetic computational models. Connectionism, which strongly emphasizes the role of learning in the creation of internal representations, has provided mechanistic models and explanations for numerous developmental phenomena. Traditional formulations of connectionism need to be made more compatible with the design of embodied systems and the central role of growth, exploration and task-independence in learning and development. We need, in addition, new connectionist modelling approaches that can deal with complex non-stationary environments and incorporate the fundamental tendency of embodied systems to act on and change their environment through sensory-motor interactions. Neuro-robotic models incorporating a blend of connectionist techniques and models that incorporate neurobiological structures and learning rules may serve as new research tools for developmental psychologists and neuroscientists interested in the behaviour of organisms from an integrated systems perspective.

Acknowledgements

Part of the work described in this chapter was supported by NIH/NIDA grant R21DA15647.

Glossary

Neuro-robotic model A model consisting of a neural simulation that is embedded in or connected to a robot. Such a model may be physically instantiated, or it may be implemented as a simulation.

Neuromodulation The functional effects of a set of neurotransmitters (called neuromodulators) on neural circuits and systems. These effects generally lead to changes in the information-processing characteristics of these circuits, for example by altering the responsiveness of their neuronal components or the efficacy of neural connections.

Phasic signal A signal that is emitted in the form of a sharply rising and falling 'spike' or 'burst'. A phasic dopamine signal consists of a temporary elevation of the discharge of dopamine neurons lasting a few hundred milliseconds.

Reinforcement learning The learning of a task, a representation or an algorithm based on reinforcement. Reinforcement refers to response-coupled events that result in the increased occurrence of the response in the future. For example, the tendency to produce an action that is reliably followed by a reward tends to be strengthened.

Temporal difference learning The learning of a task using a method for estimating the future expected reinforcement or reward. Errors in the prediction of future reward drive the learning process.

Ternary learning rule A synaptic learning rule that includes three factors: the presynaptic activity, the postsynaptic activity, and a value (or reinforcement) signal. A typical Hebbian learning rule only has two factors, the pre- and postsynaptic activities.

References

Adams, B, Breazeal, C, Brooks, R and Scassellati, B (2000). Humanoid robots: A new kind of tool. *IEEE Intelligent Systems*, 15, 25–31.

Alexander, WH and Sporns, O (2003). An embodied model of learning, plasticity and reward. *Adaptive Behavior*, 10, 141–159.

Almassy, N Edelman, GM and Sporns, O (1998). Behavioral constraints in the development of neuronal properties: A cortical model embedded in a real world device. *Cerebral Cortex*, 8, 346–361.

Angulo-Kinzler, RM (2001). Exploration and selection of intralimb coordination patterns in 3-month-old infants. *J Motor Behavior*, 33, 363–376.

Bao, S, Chan, VT and Merzenich, MM (2001). Cortical remodeling induced by activity of ventral tegmental dopamine neurons. *Nature*, 412, 79–83.

Beer, RD (2000). Dynamical approaches to cognitive science. *Trends in Cognitive Sciences*, 4, 91–99.

Beer, RD, Chiel, HJ, Quinn, RD and Ritzmann, RE (1998). Biorobotic approaches to the study of motor systems. *Current Opinion in Neurobiology*, 8, 777–782.

Berke, JD and Hyman, SE (2000). Addiction, dopamine, and the molecular mechanisms of memory. *Neuron*, 25, 515–532.

Bonci, A and Malenka, RC (1999). Properties and plasticity of excitatory synapses on dapaminergic and GABAergic cells in the ventral tegmental area. *J Neurosci*, 19, 3723–3730.

Breazeal, C (1998). A motivation system for regulating human-robot interaction. In *Proceedings of the fifteenth national conference on artificial intelligence (AAAI 98)*, pp. 54–61. Madison, WI.

Breazeal, C, Edsinger, A, Fitzpatrick, P and Scassellati, B (2001). Active vision for sociable robots. *IEEE Transactions Man Cybernetics Systems*, 31, 443–453.

Breazeal, CL (2002). *Designing sociable robots.* Cambridge, MA: MIT Press.

Edelman, GM, Reeke, GN Jr, Gall, WE, Tononi, G, Williams, D and Sporns, O (1992). Synthetic neural modeling applied to a real-world artifact. *Proceedings of the National Academy of Sciences, USA*, 89, 7267–7271.

Fellous, J-M and Linster, C (1998). Computational models of neuromodulation. *Neural Computation*, 10, 771–805.

Fitzpatrick, P and Metta, G (2003). Grounding vision through experimental manipulation. *Phil Trans R Soc Lond A*, 361, 2165–2185.

Fitzpatrick, P, Metta, G, Natale, L, Rao, S and Sandini, G (2003). Learning about objects through action—initial steps towards artificial cognition. *Proceedings of the 2003 IEEE International Conference on Robotics and Automation (ICRA)*, pp. 3140–3145.

Friston, KJ, Tononi, G, Reeke, GN, Jr, Sporns, O and Edelman, GM (1994). Value-dependent selection in the brain: Simulation in a synthetic neural model. *Neuroscience*, 59, 229–243.

Hara, F and Pfeifer, R (2003). *Morpho-functional machines: The new species—designing embodied intelligence.* Tokyo: Springer.

Hasselmo, ME (1995). Neuromodulation and cortical function: Modeling the physiological basis of behavior. *Behavioral Brain Resarch*, 67, 1–27.

Hasselmo, ME, Wyble, BP and Fransen, E (2002). Neuromodulation in mammalian nervous systems. In M Arbib (ed.) *Handbook of brain theory and neural networks, 2nd edn*, pp. 761–765. Cambridge, MA: MIT Press.

Huang, X and Weng, J (2002). Novelty and reinforcement learning in the value system of developmental robots. In Prince, Christopher G. and Demiris, Yiannis and Marom, Yuval and Kozima, Hideki and Balkenius, Christian, (eds.) *Proceedings of the second international workshop on epigenetic robotics: modeling cognitive development in robotic systems* 94, pp. 47–55. Edinburgh: Scotland.

Hyman, SE and Malenka, RC (2001). Addiction and the brain: The neurobiology of compulsion and its persistence. *Nature Reviews Neuroscience*, 2, 695–703.

Johnson, MH (2001). Functional brain development in humans. *Nature Reviews Neuroscience*, 2, 475–483.

Johnson, MH and Mareschal, D (2001). Cognitive and perceptual development during infancy. *Current Opinion in Neurobiology*, 11, 213–218.

Karmiloff-Smith, A (1998). Development itself is the key to understanding developmental disorders. *Trends Cog. Sci*, 2, 389–398.

Katz, LC and Shatz, CJ (1996). Synaptic activity and the construction of cortical circuits. *Science*, 274, 1133–1138.

Kilgard, MP, Pandya, PK, Vazquez, J, Gehi, A, Schreiner, CE and Merzenich, MM (2001). Sensory input directs spatial and temporal plasticity in primary auditory cortex. *J Neurophysiol*, 86, 326–338.

Krichmar, JL, Snook, JA, Edelman, GM and Sporns, O (2000). Experience-dependent perceptual categorization in a behaving real-world device. In JA Meyer, A Berthoz, D Floreano, H Roitblat, SW Wilson (eds) *Animals to animats 6: proceedings of the sixth* *international conference on the simulation of adaptive behavior*, pp. 41–50. Cambridge, MA: MIT Press.

Lungarella, M and Pfeifer, R (2001). Robots as cognitive tools: Information-theoretic analysis of sensory-motor data. In *Proceedings of the 2001 IEEE-RAS international conference on humanoid robots*, pp. 245–252.

Lungarella, M, Pegors, T, Bulwinkle, D, and Sporns, O (2005). Methods for quantifying the informational structure of sensory and motor data. *Neuroinformatics*, 3, 243–262.

Mareschal, D and Shultz, TR (1996). Generative connectionist networks and constructivist cognitive development. *Cognitive Development*, 11, 571–603.

Maris, M, and te Boekhorst, R (1996). Exploiting physical constraints: Heap formation through behavioral error in a group of robots. In M Asada (ed.) *Proceedings of IROS 96, IEEE/RSJ international conference on intelligent robots and systems*, pp. 1655–1660.

Montague, PR, Dayan, P and Sejnowski, TJ (1996). A framework for mesencephalic dopamine systems based on predictive hebbian learning. *Journal of Neuroscience*, 16, 1936–1947.

Nolfi, S and Parisi, D (1993). Self-selection of input stimuli for improving performance. In GA Bekey and KY Golberg (eds) *Neural networks in robotics*, pp. 403–418. Kluwer: Boston.

Nudo, R, Milliken, G, Jenkins, W and Merzenich, M (1996). Use-dependent alterations of movement representations in primary motor cortex of adult squirrel monkeys. *J Neurosci*, 16, 785–807.

Overton, PG, Richards, CD, Berry, MS and Clark, D (1999). Long-term potentiation at excitatory amino acid synapses on midbrain dopamine neurons. *Neuroreport*, 10, 221–226.

Paus, T, Zijdenbos, A, Worsley, K, Collins, DL, Blumenthal, J, Giedd, JN, Rapoport, JL and Evans, AC (1999). Structural maturation of neural pathways in children and adolescents: *In vivo* study. *Science*, 283, 1908–1911.

Pfeifer, R and Scheier, C (1999). *Understanding intelligence*. Cambridge, MA: MIT Press.

Quartz, SR (1993). Neural networks, nativism, and the plausibility of constructivism. *Cognition*, 48, 223–242.

Reeke, GN Jr and Sporns, O (1993). Behaviorally based modeling and computational approaches to neuroscience. *Ann Rev Neuroscience*, 16, 597–623.

Reeke, GN Jr, Sporns, O and Edelman, GM (1990). Synthetic neural modeling: The 'Darwin' series of recognition automata. Proc. IEEE 78, 1498–1530.

Scheier, C and Lambrinos, D (1996). Categorization in a real-world agent using haptic exploration and active perception. *Proc SAB96*, 65–74.

Scheier, C, Pfeifer, R and Kuniyoshi, Y (1998). Embedded neural networks: exploiting constraints. *Neural Networks*, 11, 1551–1569.

Schlesinger, M (2002). A lesson from robotics: Modeling infants as autonomous agents. In CG Prince, Y Demiris, Y Marom, H Kozima and C Balkenius (eds) *Proceedings of the second international workshop on epigenetic robotics: Modeling cognitive development in robotic systems*, pp. 133–140. Sweden: Lund University Cognitive Studies.

Schultz, W (1998). Predictive reward signal of dopamine neurons. *Journal of Neurophysiology*, 80, 1–27.

Schultz, W, Dayan, P and Montague, PR (1997). A neural substrate of prediction and reward. *Science*, 275, 1593–1599.

Shultz, TR (2003). *Computational developmental psychology*. Cambridge, MA: MIT Press.

Simoncelli, E and Olshausen, B (2001). Natural image statistics and neural representation. *Annual Review of Neuroscience*, 24, 1193–1215.

Singer, W (1995). Development and plasticity of cortical processing architectures. *Science*, 270, 758–764.

Sporns, O, Chialvo, D, Kaiser, M, and Hilgetag, CC (2004). Organization, development and function of complex brain networks. *Trends in Cognitive Sciences*, 8, 418–425.

Sporns, O (2002). Embodied cognition. In M. Arbib (ed.) *MIT handbook of brain theory and neural networks*, pp. 395–398. Cambridge, MA: MIT Press.

Sporns, O and Alexander, WH (2002). Neuromodulation and plasticity in an autonomous robot. *Neural Networks*, 15, 761–774.

Sporns, O and Edelman, GM (1993). Solving Bernstein's problem: A proposal for the development of coordinated movement by selection. *Child Development*, 64, 960–981.

Sporns, O and Pegors, T (2003). Generating structure in sensory data through coordinated motor activity. *Proceedings IJCNN 2003*, 2796.

Sporns, O and Pegors, T (2004). Information-theoretical aspects of embodied artificial intelligence. In: Embodied Artificial Intelligence, Iida, F., Pfeifer, R., Steels, L., and Kuniyoshi, Y. (eds.), pp. 74–85. Springer-Verlag: Berlin.

Sporns, O, Almassy, N and Edelman, GM (2000). Plasticity in value systems and its role in adaptive behavior. *Adaptive Behavior*, 8, 129–148.

Sur, M and Leamey, CA (2001). Development and plasticity of cortical areas and networks. *Nature Reviews Neuroscience*, 2, 251–262.

Suri, RE (2002). TD models of reward predictive responses in dopamine neurons. *Neural Networks*, 15, 523–533.

Suri, RE and Schultz, W (2001). Temporal difference model reproduces anticipatory neural activity. *Neural Computation*, 13, 841–862.

Thelen, E and Smith, LB (1994). *A dynamic systems approach to the development of cognition and action*. Cambridge, MA: MIT Press.

Thomas, MJ, Malenka, RC and Bonci, A (2000). Modulation of long-term depression by dopamine in the mesolimbic system. *J Neurosci*, 20, 5581–5586.

Tononi, G and Sporns, O (2003). Measuring information integration. *BMC Neuroscience*, 4, 31.

Tononi, G, Edelman, GM and Sporns, O (1998). Complexity and coherency: Integrating information in the brain. *Trends in Cognitive Sciences*, 2, 474–484.

Ungless, MA, Whistler, JL, Malenka, RC and Bonci, A (2001). Single cocaine exposure *in vivo* induces long-term potentiation in dopamine neurons. *Nature*, 411, 583–587.

Vorel, SR, Liu, X, Hayes, RJ, Spector, JA and Gardner, EL (2001). Relapse to cocaine-seeking after hippocampal theta burst stimulation. *Science*, 292, 1175–1178.

Weng, J, McClelland, J, Pentland, A, Sporns, O, Stockman, I, Sur, M, Thelen, E (2001). Autonomous mental development by robots and animals. *Science*, 291, 599–600.

Wise, RA (1996). Addictive drugs and brain stimulation reward. *Annual Reviews of Neuroscience*, 19, 319–340.

Ziemke, T (2004). Embodied AI as science: Models of embodied cognition, embodied models of cognition, or both? In:F. Iida *et al.* (eds.), Embodied artificial intelligence, Lecture Notes in Artificial Intelligence 3139, pp. 27–36. Springer: Berlin.

Chapter 9

Phonological deficits and developmental language impairments: evidence from connectionist models

Marc F. Joanisse

Introduction

Children with Specific Language Impairment (SLI) fail to acquire age-appropriate language skills despite having no obvious sensory, cognitive or neurological deficits. The disorder has been identified in approximately five per cent of children, and is typically diagnosed as a significant limitation in language comprehension and production, in spite of normal sensory, cognitive and neurological capacities (Bishop, 1997; Leonard, 1998). Children with SLI learn to speak much later than their peers, produce fewer and shorter utterances than expected for their age and intelligence, and have difficulties with elements of language such as phonology, morphology and syntax.

Much of the interest in SLI has come from what it can teach us about the cognitive bases of intact language abilities. As research in language acquisition continues to try to come to terms with how children acquire complex language skills in an apparently effortless fashion, SLI can be seen as a corollary case of why some children find language acquisition a significantly difficult task. A well-studied example of this is affected children's poor production and comprehension of grammatical morphology, most famously past tenses in English (Gopnik, 1997; Marchman *et al.*, 1999; van der Lely and Ullman, 2001). Theories of formal grammar characterize morphological systems is as sets of symbolic rules; for instance, past tense formation can be captured with rules like '*add –ed*.' Curiously, children with SLI tend to lag behind in their development of past tense, for instance producing many errors of morphological omission (e.g., **yesterday he bake a cookie*). This has led some to the conclusion that SLI represents a deficit in learning and representing grammatical rules (Pinker, 1991; Gopnik, 1997; van der Lely and Ullman, 2001).

A second element of language that is commonly impaired in SLI is syntax, the knowledge of sentence structure. Understanding and producing sentences requires language users to recognize configural aspects of sentences that extend beyond the linear order of words (Chomsky, 1981). For instance, a subtle difference in structure can lead to a complete change in meaning, as in the sentences '*The cop that the robber hit ran away*' and '*The cop that hit the robber ran away.*' Like morphology, syntax is purported to be governed by complex rules that describe, among other things, subject–verb agreement, the assignment of subjects and objects to verbs, and the relationships between pronouns and their referents. Children with SLI have marked deficits in some aspects of syntax, such as comprehending reversible passives (e.g., *The boy is tickled by the girl*) (van der Lely, 1996). They are also poor at resolving 'bound' pronouns and reflexives, which refer to a noun in the same sentence (van der Lely and Stollwerck, 1997). For instance, in the sentences '*Robbie likes when William tickles him,*' and '*Robbie likes when William tickles himself*' children with SLI have difficulty determining who *him* and *himself* refer to. Here again, one might characterize these difficulties as evidence for a deficit in learning syntactic rules or principles (van der Lely and Stollwerck, 1997).

In the present work, I argue that the characterization of SLI as a grammar-specific deficit is an over-simplification. Instead, I suggest that SLI is caused by these children's problems with speech perception. These speech perception deficits impair the development of phonological representations, which are in turn the proximal cause of language deficits in these children (Chiat, 2001; Evans, 2001). I argue that the aspects of language that these children find difficult, along with the way in which they are impaired, can give us useful insights to understanding the cognitive and perceptual prerequisites for language acquisition in typically developing children. To illustrate this point, I present the results of two connectionist simulations that illustrate how a basic speech perception deficit can lead to the specific types of problems that these children have with morphology and syntax.

The perceptual deficit theory has been highly disputed in the SLI literature on the basis of whether the grammar deficits that are observed in these children can adequately be explained by a perception deficit; simply put, it can be difficult to imagine how a basic speech processing deficit can lead to the specific constellation of grammar problems in these children. Connectionist simulations have the potential to inform this debate by allowing the researcher to test a skill such as language processing within a computerized model and then directly investigate the conditions under which this skill develops either normally or in an impaired fashion. As such, these models are an ideal way to explicitly test the impact of a perceptual deficit on the development of phonology, morphology and syntax, in a way that can be directly compared to the empirical facts.

The perceptual basis of SLI

The hypothesis that is presented in this work is that children with SLI acquire language poorly because of their difficulty perceiving speech. This deficit has a number of consequences, but most proximally affects phonological development; it is this phonological deficit that in turn causes these children's problems with morphology and syntax. This theory grows out of earlier perception-based theories of SLI, which hold that these children's language difficulties stem from a deficit in perceiving elements of speech critical for learning phonemes and morphemes (Tallal *et al.*, 1993; Leonard, 1998). It is proposed that phonological deficits are the key link between perceptual and grammatical impairments. It is further proposed that children with SLI are not impaired on morphology and syntax strictly because they have difficulty perceiving morphemes and words, but instead because they have difficulty translating the auditory forms of words into a phonological code, maintaining this code in working memory, and using it to learn generalizations about systematicities in language.

The term 'Specific Language Impairment' is based on the definition of a grammatical language impairment that precludes cognitive, frank neurological and sensory deficits (Johnston, 1984). Thus, the perceptual deficit theory seems anathema to the SLI definition; I suggest only that the term best fits the class of children who have developmental deficits primarily in the area of grammatical language processing. Indeed, there is evidence to suggest SLI is not quite as 'specific' as has previously been thought, given the finding of non-linguistic deficits in these children (Ellis Weismer, 1985; Nelson *et al.*, 1987). While some small subset of these could appear to have deficits that are strictly localized to language (van der Lely *et al.*, 1998), I and others suggest those children represent the extreme high end of the distribution of cognitive abilities in SLI, rather than forming a distinct category of 'grammatical' SLI (Bishop *et al.*, 2000). Indeed, further testing may well reveal that such children do in fact have subtle deficits in speech perception and phonology, and that these could explain their problems with morphology and syntax.

Of primary interest to the present theory is the considerable evidence of subtle auditory and speech perception deficits in SLI. These children are poor at discriminating place of articulation and voicing features of consonants

(Sussman, 1993; Stark and Heinz, 1996; Joanisse *et al.*, 2000), indicating their categorical perception of such stimuli is impaired. For example, they are poorer than controls at discriminating and identifying voice onset time (VOT) acoustic cues, which differentiate voiced and voiceless stop consonants (e.g., the phonological feature that distinguishes *dug* from *tug*.) While VOT is an acoustically continuous feature, normally developing children and adults tend to be categorical in their perception of it (Liberman *et al.*, 1957), categorizing both end point and mid point stimuli equally well within a single speech category. In contrast, language-impaired children tend to be less categorical on such stimuli, marked by less consistency on end points and a shallower slope between the two. I argue that these types of speech problems either reflect or are the cause of poor phonological organization, which in turn impairs the development of more complex language abilities.

Several theories have been put forward to explain why speech perception is impaired in SLI. One such theory focuses on impaired processing of rapid, sequential information (Tallal *et al.*, 1993). Since speech perception involves sensitivity to a rapid and transient auditory signal, a deficit in resolving these types of signals would interfere greatly with speech processing. This theory predicts that these children will tend to be selectively impaired in perceiving acoustic cues to speech involving brief transient acoustic cues, for example VOT and formant transition cues used in identifying stop consonants. The temporal deficit theory has been challenged on the basis of findings that some children with SLI do not appear to have nonspeech perception deficits (van der Lely *et al.*, 1998; Bishop *et al.*, 1999). Non-speech perceptual deficits thus continue to be the matter of much debate in the SLI literature. Nevertheless, the theory of SLI that is presented here does not necessarily predict that these children have such a deficit. It is also possible that these children have a speech-specific deficit, either at the level of perception or perhaps at the level of phonological working memory (Gathercole and Baddeley, 1990; Briscoe *et al.*, 2001).

Indeed, a growing literature suggests that speech perception impairments *are* widespread in SLI. Using sensitive measures of speech perception that directly manipulate acoustic parameters such as VOT, many studies have found fundamental differences in how language impaired children categorize and discriminate speech sounds compared to other children (Bernstein and Stark, 1985; Stark and Heinz, 1996; Joanisse *et al.*, 2000). While other studies have suggested apparently normal auditory word identification in some children with SLI (Gopnik, 1997), such results need to be interpreted cautiously, given the types of tasks that are being used, and also the possibility that perceptual deficits early in development can resolve over time such that they can become undetectable later on, but nevertheless leave lasting problems in language abilities (Bernstein and Stark, 1985).

Assuming that speech perception deficits are widespread in SLI, the next question is how this type of deficit is related to these children's impaired grammar. It is proposed that phonology provides the link between the two. Phonology is arguably the first aspect of language that children acquire, and provides a gateway to other elements of language such as morphology and syntax. Children learn many of the phonological properties of their language at a very young age (Jusczyk, 1997). Impaired speech perception in infancy could thus interfere with the development of phonological representations from an early age, which in turn can affect the development of other aspects of language. This idea is not new, building among other things on Leonard and colleagues' 'surface hypothesis,' which maintains that children's ability to learn grammatical morphemes is impaired by difficulties in consistently perceiving them in less salient contexts (Leonard, 1998).

While phonology has not been widely studied in the SLI literature, there is clear evidence that these children do tend to have phonological problems. They have problems with expressive phonology such as misarticulating and deleting phonemes in words (Leonard, 1982). In addition, they have poor non-word repetition, particularly as word length increases, indicating phonological short-term memory problems (Kamhi *et al.*, 1988; Montgomery, 1995; Gathercole and Baddeley, 1990,). They are also poor on phonological awareness tasks: they have difficulty identifying words with similar phonemes, such as deciding whether *Sam* is more like *sock* or *ball* (Bird and Bishop, 1992); similarly, they are poor at phoneme deletion (e.g., saying the word *split* without the /l/) and other phoneme segmentation tasks (Kamhi *et al.*, 1988; Joanisse *et al.*, 2000). The phonological deficit theory holds that these speech perception and phonology problems are the direct cause of these children's grammar deficits.

Past tense deficits in SLI

Given that children with SLI have problems with speech perception and phonology, how might this deficit affect how they learn grammatical relationships such as past tense? The answer lies in the theory that morphology involves the integration of knowledge about how words are semantically and phonologically related (Joanisse and Seidenberg, 1999; Seidenberg and Gonnerman, 2000). English past tense represents a useful illustration; forming a past tense actually involves three related patterns: if a verb's final phoneme is a voiceless consonant, add /t/ (e.g., *walk—walked*); if it is a voiced consonant or a vowel, then add /d/ (e.g., *try—tried*); if it is an alveolar stop (/t/ or /d/) add

/Id/ (e.g., bait—baited). The fact that *walked, tried* and *baited* all take slightly different past tense endings (known as *allomorphs*) nicely illustrates how a phonological component is intrinsic of morphological processing. Instead of learning to add a suffix to a base form, the learner must analyse present and past tense forms and discover the conditions under which each allomorph surfaces. The crux of the phonological deficit theory is that this task is more difficult in children with impaired speech perception for at least two reasons. First, this morpheme is perceptually non-salient, because it involves stop consonants that tend to be devoiced and unreleased word-finally (Leonard, 1998). Second, ill-formed phonological representations resulting from a perceptual deficit will weaken the ability to analyze and learn subtle aspects of a morphological process, such as how alveolar and continuant phonological features influence past tense inflection.

The grammar deficit approach

Theories seeking to characterize SLI as a grammar impairment explain their past tense deficits as the result of deficits with learning grammatical rules. A prediction of this theory is that these children should have less difficulty with forms that are not rule-like. English past tense provides us with a test case, in the form of irregulars that defy these rules (e.g., *take-took* and *buy-bought*). According to a popular rule-based theory of past tense, irregular forms are encoded using a separate mechanism that is not rule-governed, and which should therefore be intact in SLI (Pinker, 1991). Contrary to this prediction however, studies have found poor performance for both regular and irregular morphology in SLI (Marchman *et al.*, 1999; van der Lely and Ullman, 2001), as illustrated in Table 9.1. Instead, children with SLI appear to be numerically *worse* on irregulars than regulars, reflecting the normal trend of delayed development of irregulars in children (Plunkett and Marchman, 1993).

Table 9.1 Past tense morphology in children with SLI, compared to typically developing children*

	SLI	Younger controls	Age-matched controls
Regulars	22	58	96
Irregulars	18	27	89
Nonwords	7	31	75

*Data are in percentages correctly inflected forms in a sentence completion task (*The boy likes to walk; yesterday he ____,*) computed across all forms used in their study. From van der Lely and Ullman (2001).

Children with SLI also fail to use the regular past tense form as productively as typically developing children, which has again been put forward as supporting a rule deficit theory. For instance, typically developing children occasionally make errors in which they over-regularize the regular past tense to irregulars (e.g., producing **taked*, **swimmed* and **sleeped* instead of *took*, *swam* and *slept*) (Marcus *et al.*, 1992; Plunkett and Marchman, 1993). These errors suggest that they have learned the regular past tense rule and are over-generalizing it to other forms. In contrast, children with SLI produce fewer errors of these type (Marchman *et al.*, 1999). Another useful measure of this are nonword forms, since these measure children's ability to creatively apply past tenses to unfamiliar cases (Berko, 1958). Children with SLI tend to have marked difficulty generating past tenses of nonwords like *blick* and *trog*, compared to normally developing children (van der Lely and Ullman, 2001). Both these findings have led some to suggest these children are not creatively using the regular past tense rule the way typically developing children do (Gopnik and Goad, 1997; van der Lely and Ullman, 2001).

On closer examination however, the data fail to uphold this explanation. First, it is not the case that children with SLI *never* produce over-regularizations; they simply produce fewer of them than their typically developing peers (Marchman *et al.*, 1999). These children are also capable of generalizing the regular ending to nonwords, but not to the same extent as other children (e.g., Table 9.1). This suggests these children *do* have some degree of proficiency with the regular ending, contrary to the predictions of a grammar deficit theory. Instead, these data suggest that a more general processing deficit is interfering with learning and using past tenses, as I illustrate below.

Modelling the phonological deficit theory

The alternative explanation for past tense deficits in SLI is that these individuals have difficulty processing phonological information, which in turn leads to problems learning the consistency in the relationship between present and past tense forms (e.g., the difference between *walk* and *walked*) and the commonalities amongst regular past tenses (e.g., the overlap between *walked*, *marched*, *flipped* and *flashed*). Though there is evidence suggesting children with SLI have problems with speech perception, phonology and morphology (Joanisse *et al.*, 2000), what is missing is an illustration of precisely how these phenomena are linked. To address this, the theory has been implemented in a connectionist model that simulates the effect of a perceptual deficit on various aspects of past tense learning (Joanisse, 2000, Joanisse, 2004). This model learned to generate past tenses of verbs in the process of learning relationships between the meanings (semantics) and sounds (phonology) of verbs. A developmental impairment was simulated by impairing how the network perceived phonological information, consistent with what is theorized to occur in a speech perception disorder.

The model builds on two earlier simulations of past tense deficits. In an earlier model, Hoeffer and McClelland (1993) examined a perceptual impairment in learning past tense. Their model took the semantics of a verb on the input and output its phonological form. They simulated the perceptual weakness of wordfinal stops by reducing the strength of connections to word-final phonemes in the output layer (e.g., the /t/ in past tenses like *snapped*, but also in present tenses like *blast*). They found that this reduced salience did not affect their network's ability to learn past tenses under normal circumstances, because some perceptual information was still available to it. However, when they further degraded the input to the phonological layer as a whole, simulating generally poor speech perception in SLI, they found the network had significant difficulty learning the past tense rule. For instance, it was very poor at generating a past tense like *typed*. On the other hand, it was much better at producing words that were phonologically similar to past tenses, like *fact*. The authors argued that this was due to competition between related present and past tense forms such as *type—typed*, which share a large degree of semantic and phonological overlap.

The Hoeffner and McClelland model shows how a speech perception deficit can disproportionately affect past tense processing, due to competition between the phonological forms of a word's present and past tense. Their simulation also illustrates a language impairment that affects morphological knowledge in a partial, rather than wholesale manner. As in SLI, their model did produce some appropriate past tense forms, but to a lesser degree than what is seen in normal development. This is an important difference between connectionist and rule-based theories of SLI. Under the latter theory, the mechanism by which a past tense rule can be impaired while preserving *some* past tense forms is unclear. For instance, a rule deficit theory does not predict a mechanism by which a child with SLI will produce over-regularizations and past tenses of nonwords (which they *can* do, to a lesser extent than their peers). The phonological deficit account holds that language impaired children's knowledge of a morphological alternation is imperfect, rather than absent altogether, explaining why they do show some generalization ability.

The present model also builds on the Joanisse and Seidenberg (1999) model of past tense deficits in aphasic adults with damage to either Broca's area or regions of the basal ganglia projecting onto it (Ullman *et al.*, 1997). The model

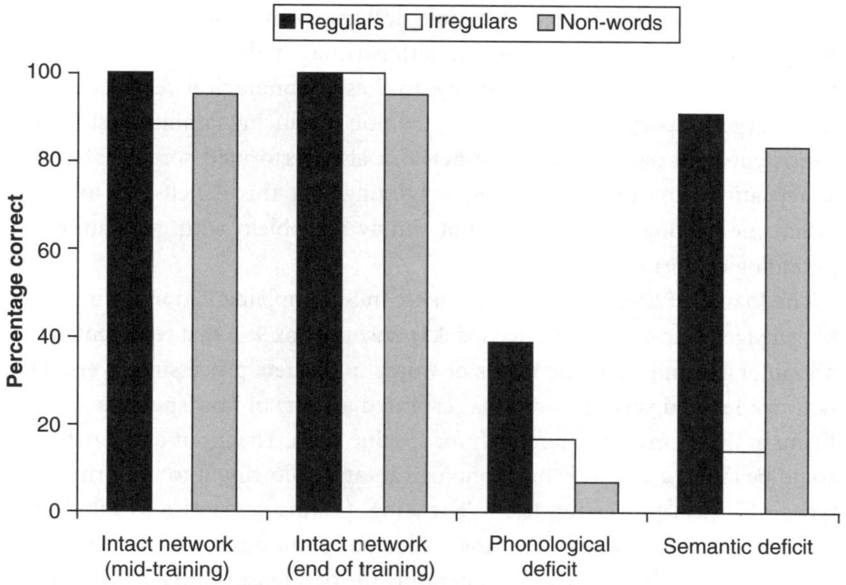

Fig. 9.1 Data from connectionist models of normal and impaired past tense acquisition. The intact network showed slower development of irregulars compared to regulars and non-words over the course of training. In contrast, a phonological impairment yielded SLI-like deficits characterized by delayed overall acquisition, and very poor non-word generalization. A semantic deficit resulted in a different pattern of impairment marked by poor performance on irregulars.

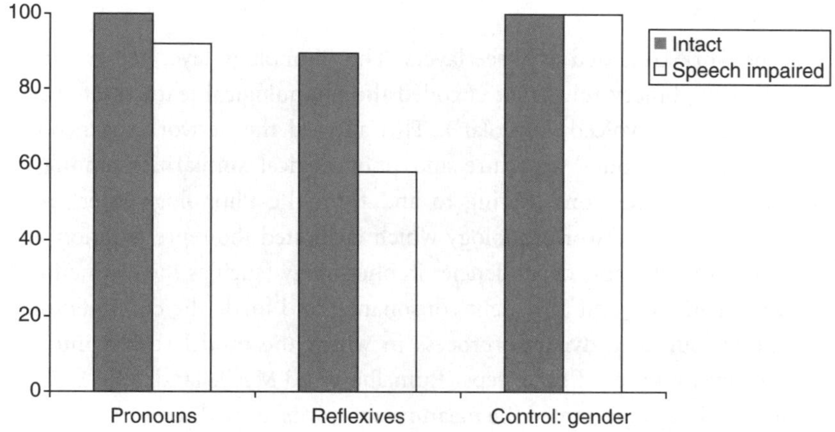

Fig. 9.2 Comparison of performance of the intact and phonologically impaired sentence comprehension networks on resolving bound pronouns and reflexives, compared to a control condition in which pronouns could be resolved using only gender information. The results indicate poorer than normal performance in the speech impaired network on sentences requiring the use of syntactic binding, but better performance when gender information could also be used.

explored the possibility that a phonological impairment would lead to poor morphological generalization by simulating damage to brain areas responsible for phonological processing. Damage to these phonological representations had a larger impact on generalization than on producing familiar past tenses, consistent with patient data. The network also performed somewhat better at repeating nonword past tenses, suggesting that this deficit was to some extent morphology-specific, and not simply a problem with generating the phonological forms of words.

The Joanisse (2000) model of past tense in SLI implemented morphological knowledge using a recurrent (feedback) network (Box 9.1) that represented the phonological and semantic forms of words in discrete processing layers. The network learned verbs in ways that captured aspects of how speakers acquire them: in the course of using words for specific tasks. The input to the network could be either a semantic or phonological form, allowing it to perform three types of lexical processing tasks: 'Speaking' (taking semantics as input and outputting its phonology) 'Hearing' (taking a sound input and activating the semantics) and 'Transforming' (generating the past tense given a phonological present tense form as an input).

Box 9.1 The Joanisse (2000) model of past tense impairments in SLI

Structure

The network consisted of three layers. The Phonology layer represented words using binary units that encoded the phonological features of each phoneme (e.g., 'voiced', 'alveolar'). This allowed the network to encode words' internal sound structure and phonological similarities amongst words. The connections leading to and from the Phonology layer created a recurrent network topology which facilitated the representation of non-linearly separable dependencies in phonology (such as the distinction between sonorant and obstruent consonants) and made the computation of speech output a dynamic process in which the model settled into a pattern over a series of time steps (Rumelhart and McClelland, 1981). The Semantics layer represented the meanings of words using discrete units that signaled individual word concepts (e.g., the verb *take* activated the [TAKE] concept node). In addition, a separate unit signaled the past tense concept (e.g., *took* activated the [TAKE] and [PAST TENSE] nodes). These layers were interconnected through a hidden layer consisting of 150 units, the purpose of which was to increase the network's representational capacity

Box 9.1 *(continued)*

by providing it with a set of units that passed activations to and from the phonology and semantics layers.

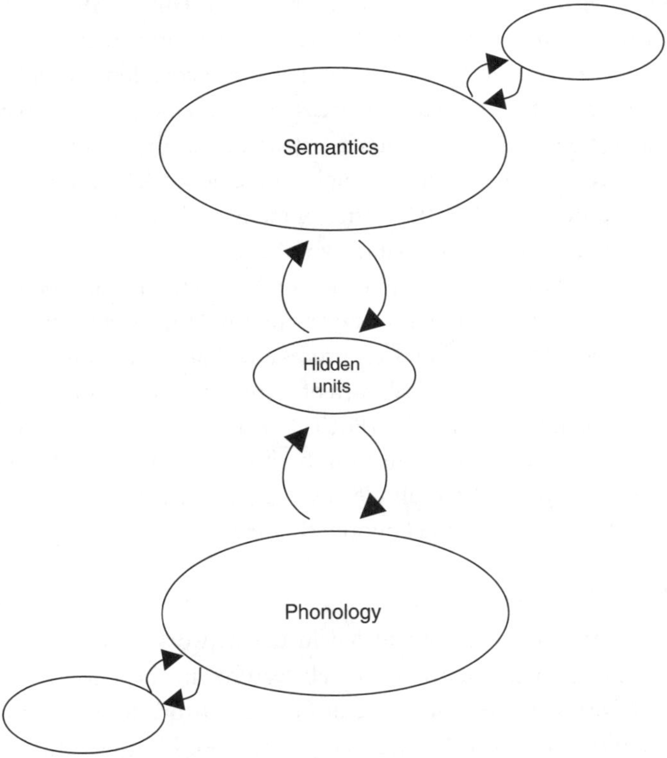

Fig. 9.3 Network article used in Joanisse (2000) to simulate learning of past tense morphology.

Learning

The network learned a vocabulary of 600 monosyllabic verbs in English (two-syllable past tenses such as *waited* were also included in training.) These verbs were frequency weighted such that the network was exposed to high frequency verbs (e.g., *say-said*) more frequently than lower frequency verbs (e.g., *sail-sailed*). Three task types were used in training: 'Speaking' trials involved activating the semantic representation of a word and producing its phonological form on the phonology layer; 'Hearing' involved taking the word's phonological form and activating its representation on the Semantics LAYER; and 'Transforming' involved taking a present tense verb's phonological form, activating the past tense node on the Semantics layer and generating the verb's past tense on the phonology layer. Because

Box 9.1 *(continued)*

exposure to these tasks was interleaved, the model had to find a set of connection weights that would allow it to perform all of these tasks accurately.

The back-propagation through time algorithm (Hinton, 1989), was used to train the network. In this mechanism, the network maps an input to an output as a dynamical process of propagating activation throughout the network over a set of discrete time steps. Next, the network compares the target activation values to the actual activations of the output units, and adjusts the connection weights throughout the network in a way that minimizes this difference. Over the course of many trials, the network achieves optimal performance on the training vocabulary.

The recurrent (feedback) architecture of the model allowed activation to propagate throughout the network over multiple time steps, which simulated dynamical processes in lexical processing. For instance, in the 'Transforming' trials, the phonological form of a present tense verb was presented to the Phonology layer, and the [PAST TENSE] unit was activated in the Semantics layer. Next, activation from both layers propagated throughout the network's layers over multiple time steps, ultimately generating the correct past tense form in the Phonology layer.

Simulating a deficit

A speech impairment was simulated in the network by adding random Gaussian noise to the activation of all features in the Phonology layer. This effectively made it more difficult for the network to build consistent representations of words due to small perturbations in their phonological forms from one instance to the next.

When trained, the network achieved near adult-like ability on the items in the training set, and was able to accurately transform nonwords like *stup* into past tenses like *stupped.* Next, a second network was trained, but in this network a speech perception deficit was simulated by adding random noise to the features in the Phonology layer. This deficit made it difficult for the network to accurately encode phonological forms, simulating how a speech perception deficit might influence word learning. As expected, this speech impaired model's performance at the end of training was significantly poorer than that of the intact network. It failed to learn many of the items in the training set and performed much worse than the intact network on the three training tasks.

To more directly assess how performance of the speech impaired and intact networks differed, the intact and speech impaired networks were compared on the transformation task using regular, irregular and nonword verbs that have previously been used to test children with SLI and adult aphasics (Ullman *et al.*, 1997; van der Lely and Ullman, 2001). Figure 9.1 illustrates how the two networks differed in processing these different item types, and reveals some interesting patterns of performance. The intact model was tested at two points mid-training, to better assess normal development in the network. It seemed to be learning regulars more quickly than irregulars, similar to what is observed in normally developing children. The speech impaired network also showed this same delay for irregulars, which seems to reflect what is observed in SLI. Its performance on non-words was quite different however; whereas non-word performance was quite good for the intact network, the speech impaired network achieved only a very poor degree of accuracy on these forms, suggesting it was not able to generalize the past tense ending to unfamiliar forms.

The speech impaired network also produced a different pattern of errors than the intact network, also consistent with SLI: it produced many fewer over-generalization errors on irregular verbs (e.g., **taked*, **sleeped*), but a greater rate of zero-marking errors (e.g., errors of omission such as producing *take* as the past tense of *take*). The speech impaired network's error patterns suggest that it was not acquiring familiar past tenses in a way that reflected a simple delayed pattern. Instead, the results indicate that a perceptual deficit led to a deviant pattern of past tense acquisition in this network. The results further suggest that this developmental profile need not be characterized as the result of a deficit in learning grammatical rules, and can instead be explained as resulting from a deficit in perceiving speech.

One question is whether the pattern of deficits that the network produced is attributable to the network's phonological problems, or could have resulted from some other type of deficit as well. This was tested by training a new network in which a semantic deficit was simulated by adding random noise to the Semantics layer during training. This network also demonstrated slower learning of verbs, and failed to reach the same level of performance as the intact network. However, the pattern of impairment that the network yielded was quite different. As illustrated in Figure 9.1, the semantic deficit resulted in very poor performance on irregulars compared to both regulars and nonwords; further analyses also revealed that this network generated a pattern of over-regularization and zero-marking errors that was much more similar to the intact network (Joanisse, 2000, Joanisse, 2004). The results of this simulation indicated that the way in which a network is impaired directly influences the types of deficits that it will demonstrate. The speech deficit tended to affect generalization of past tenses to nonwords, similar to what is observed in SLI. In contrast, a semantic deficit yielded a pattern consistent with delayed development, marked by poorer than expected performance in irregulars.

Discussion

The model of past tense presented here illustrates how morphology can be acquired within a mechanism that learns to map words' phonological and semantic forms. They represent a radically different approach to morphology than what is typical of generative linguistics, which views morphological rules as functionally and neurally separate from the mechanism responsible for processing exceptional forms (Pinker, 1991; Marcus *et al.*, 1992; Ullman *et al.*, 1997). This 'Words and Rules' account explains morphological acquisition as two parallel tasks: inferring a set of symbolic rules that capture the default regulars, and memorizing irregulars as discrete lexicalized forms. By this same account, SLI is viewed as a deficit in processing morphological rules, such that generalization is impaired while memorized forms are left intact.

The present work illustrates some clear drawbacks of this approach. First, children with SLI do not perform worse on regulars than irregulars; if anything, the dissociation appears to be the opposite, marked by worse performance on irregulars than regulars. Second, rule deficit theories ignore the fact that these children's morphological deficits coincide with more general processing impairments, for instance speech perception difficulties. I argue that the convergence of codes theory of morphology can better capture morphology problems in SLI, by explaining these children's deficits as resulting from phonological factors. One thing that distinguishes this approach from the Words and Rules theory is that it does not posit discrete routes or mechanisms for encoding regular and irregular forms. Although I have implicated multiple types of knowledge in morphological processing, differences in how forms are computed are due to the degree to which they rely on these knowledge types. In the network, rule-like behaviour is promoted by phonology because of how it allows the network to generalize across forms based on phonological similarity. Learning idiosyncratic forms is supported by weights connecting semantics and phonology, which allow the network to override the generalization process when an irregular form is encountered.

This model can also explain impaired morphology in SLI. Consider how the network learned past tenses when words' phonological forms were distorted, as would occur in the case of a speech perception deficit. Under these circumstances, the network had a reduced ability to compute phonological similarities between present and past tense forms of words, and across all past tense forms. This was due to the fact that a word's phonological form was different each time it was presented to the network, making it difficult to learn generalizations about past tense morphology. An important observation about the resulting deficit is that the network was not specifically impaired in any single aspect of morphology; the speech impaired network performed poorly on all types of forms relative to the intact network. However, nonwords were the most impaired because they relied heavily on phonological knowledge to transform them into its past tense, whereas familiar forms were less impaired thanks to the network's word-specific representations of these forms.

Syntax deficits

The results of the model presented above speak to the theory that language impairments occur due to problems with more basic knowledge such as phonology and semantics. Next I turn to a similar model that tested how this same theory can explain syntactic comprehension deficits in SLI (Joanisse and Seidenberg, 2003). These children have difficulty with configural aspects of sentences, such as determining how anaphoric pronouns are bound to their antecedents and how active voice sentences are transformed into passives. It can be difficult to imagine how subtle impairments in processing speech can cause problems with syntax. The benefit of a modeling approach is the ability to illustrate *in machina* how this might occur.

Children with SLI have been shown to have difficulty resolving the antecedents of pronouns and reflexives in sentences. In generative syntax, the relationship between pronouns and the nouns they refer to is governed by *binding conditions* (Chomsky, 1981). For instance, binding conditions are used to capture why *himself* refers to *Stan* in the first sentence below, but cannot refer to *Stan* the second sentence:

Stan likes to talk about himself
Stan's best friend likes to talk about himself

Generative accounts hold that learning the difference between the two involves recognizing structural properties of sentences that extend beyond linear order (e.g., Chien and Wexler, 1990). Since children with SLI have difficulties with these types of sentences, they are claimed to have a specific deficit in representing syntactic information (van der Lely and Stollwerck, 1997).

The alternative account of this is that syntax in SLI is much more similar to that of normally developing children that has been previously claimed. Instead, the deficit reflects a processing impairment that limits the use of syntactic knowledge. This account turns on the theory that a speech perception deficit leads to reduced phonological working memory capacity. Because normal sentence comprehension relies on working memory to resolve some syntactic relations (Daneman and Carpenter, 1983), a working memory impairment leads to poor syntactic processing. A connectionist model of syntax is presented that tests the theory that impaired phonology in SLI leads to poor sentence comprehension, due to the role of working memory (Gathercole and Baddeley, 1990; Montgomery, 1995).

Pronoun deficits in SLI: syntax or working memory?

Pronoun processing in SLI has been tested using a sentence-picture matching task in which children compared pictures to sentences like *Every monkey is tickling himself*, (van der Lely and Stollwerck, 1997). Since pronouns can sometimes be resolved non-syntactically using pragmatic and contextual knowledge, children were also tested on sentences in which the antecedent could be inferred strictly from gender information. For example, *himself* in *Shirley thinks Stan likes himself* can be resolved as referring to *Stan* based on the fact that only one male entity is mentioned in the sentence. The study found that children with SLI were poor on sentences in which pronouns and reflexives could only be resolved syntactically. However, children were much better when pronoun reference could be inferred using contextual gender information. The authors concluded that SLI involves a grammar-specific deficit in processing syntactic structure but leaves lexical and pragmatic knowledge intact.

One complication to this conclusion involves the extent to which SLI children performed at chance on sentences requiring the use of syntactic knowledge. While this was the case on some conditions, the same children were at or near ceiling on other sentences that also required syntax to resolve pronouns. When actual detection rates are considered within conditions (hits + correct rejections/false alarms + incorrect rejections), SLI subjects can be characterized as responding better than chance on all conditions (Bishop, 1997). This seems to suggest that these children are using some amount of syntactic information in this task, but are nevertheless not as proficient as controls.

The alternative view of syntactic deficits in SLI is that these children *are* in fact able to process syntactic relationships, but that a processing deficit is hindering their ability to do so. On this view, a perceptual deficit is responsible for sentence comprehension deficits, due its role in phonological processing and working memory. Sentence recognition typically involves maintaining the phonological form of a sentence in working memory in order to determine the critical relationships between its constituents and to resolve ambiguities. Accordingly, individual differences in working memory span influence processing syntactically complex sentences and pronouns (Daneman and Carpenter, 1983; King and Just, 1991). Similarly, the quality of the phonological code that is used to maintain sentences in working memory also affects sentence processing; for example, sentences containing many words with the same initial phoneme (similar to tongue twisters) tend to be more difficult to comprehend (McCutchen *et al.*, 1994). In addition, individuals have greater difficulty processing auditory sentences when they are acoustically distorted or embedded in noise (Dick *et al.*, 2001).

It is proposed that SLI represents an extreme case of this, in which a weakened phonological code adversely influences working memory capacity, which in turn leads to sentence comprehension deficits. On this account, speech perception deficits lead to impaired phonological representations that in turn have a negative impact on the capacity to maintain sentences in working memory. This is supported by a growing body of research showing language impaired children have problems on many tasks that draw on short-term memory, and that this problem is strongly correlated with the severity of their syntactic processing difficulties (Montgomery, 1995).

Simulating the acquisition of bound pronouns

Pronoun resolution deficits in SLI were modelled in a network architecture that simulated some basic aspects of sentence comprehension (Joanisse and Seidenberg, 2003). The network learned to recognize words presented in sentence contexts. To perform this task accurately, the network had to discover syntactic structure. Some of the sentences also contained bound pronouns, which required the network to resolve long-distance syntactic dependencies. The architecture of this network is illustrated in Box 9.2, and is very similar to the one used to model morphology, except that it contained a simulated working memory resource within the network's hidden layers. The input layer encoded phonological features of words, whereas the output layer encoded words' meanings. Both phonology and semantics were represented using distributed representational schemes that allowed the network to encode the overlap in both the sounds and meanings of words; for instance, it was able to encode the phonological overlap between 'Mary' and 'Larry', and the semantic overlap between 'hit' and 'slap'.

Box 9.2 A connectionist model of sentence comprehension (Joanisse and Seidenberg, 2003)

Structure

The architecture of the model was very similar to the past tense model described in Box 9.1, except that the hidden layer consisted of two sets of 50 units each that formed a feedback loop. This loop of connections helped the network maintain an internal representation of a sentence over time. The Phonology layer encoded the sounds of words using a set of distributed phonetic features. The Semantics layer used a set of distributed semantic features that helped encode meanings of words (e.g., the concept of CAT was encoded by activating units corresponding to features such as [ANIMAL], [FELINE], [PET], [FURRY] etc.)

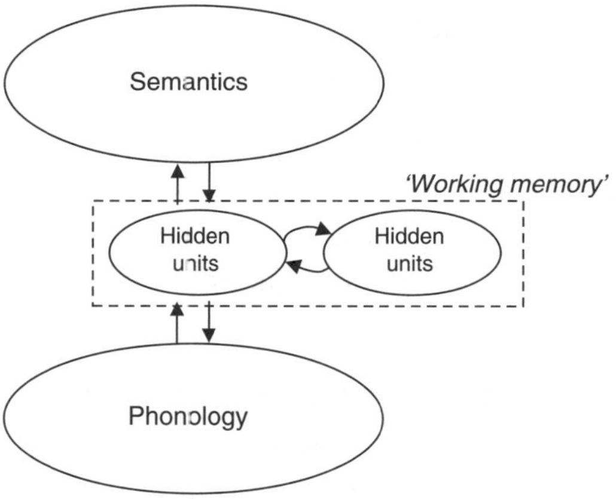

Fig. 9.4 Network architecture used in Joanisse and Seidenberg (2003) to simulate pronoun comprehension in sentences.

Learning

The network's task was to recognize words' phonological forms by activating the appropriate semantic forms. Sequences of these words formed grammatical sentences, such that on each trial the network was exposed to an entire sentence, and was able to use syntactic context in order to better recognize the meanings of words. For example, on a typical trial the network was input the words *Linda, danced, with*, and *Barney*, and its task was to activate the appropriate meanings of each word. The network was trained on 40,000 sentences, 25 per cent of which contained pronouns or reflexives. For these items, the network was required to output the semantics of the word that the pronoun referred to (e.g., for *himself* in

Box 9.2 *(continued)*

'*Bob danced with himself*', the network was trained to output the semantics of *Bob*).

Simulating a deficit

A speech impairment was simulated by adding random Gaussian noise to activations in the Phonology layer. It was observed that this noise would not affect word recognition under normal circumstances. However, this noise did tend to make it more difficult for the network to resolve pronouns, because of the impact that it had on maintaining representations of sentences within the working memory mechanism that was simulated in the hidden unit layers.

The network learned to recognize sentences by mapping sequences of phonologically encoded words to semantically encoded outputs. Each word was input one at a time to the phonology layer and the network learned to activate the meaning of each. A variety of sentence types were used in training, including sentences with embedded clauses (*Bob says Linda danced with Barney*,) and also sentences containing bound pronouns and reflexives, (*Bob danced with himself, Bob says Barney danced with him*).

Two networks were trained; one network simulated SLI through the addition of a simulated speech perception deficit, implemented by adding random noise to the phonological forms of words during training. At the end of training, the networks were compared on a number of sentence types, including a set of novel sentences that contained bound pronouns or reflexives. It was found that the intact network learned the training sentences to a high degree of accuracy, and was also performing very accurately on novel sentences that tested the resolution of pronouns and reflexives. In comparison, the speech impaired network failed to learn the training set to the same degree of accuracy, and performed much worse on the test sentences (Figure 9.2). The network's errors were similar to those of children with SLI in two major respects. First, it performed poorly on the pronoun and reflexive sentences, but was still above chance levels on all conditions. This indicated that the speech deficit was causing poor pronoun comprehension, but had not eliminated this capacity altogether. Second, this network performed much better on a third set of test sentences containing pronouns and reflexives that could be resolved using only gender information (e.g., *Bill says Mary is kicking herself*). Remarkably, it performed at approximately the same level as the intact network on this condition. This result is important because it indicates that a speech deficit did not evenly degrade all aspects of sentence processing in this network; aspects of comprehension related to recognizing individual words in context and using contextual information to resolve pronouns remained intact.

This model of SLI nicely demonstrates how a speech impairment can lead to sentence comprehension difficulties, and why this deficit is not so severe as to impair all aspects of sentence comprehension. Instead, it had the strongest impact on the ability to process pronoun reference, due to the importance of phonological information for holding sentence information in working memory. The addition of noise to the network's phonological inputs weakened its ability to maintain accurate working memory representations of words over time. However, it left intact the network's ability to use context to resolve pronouns, when such cues were available.

Conclusion

A variety of morphosyntactic deficits have been identified in language impaired children. While such deficits have been taken to indicate that a module of linguistic competence is either impaired or missing, the present work explored the alternative theory: that deficits in language learning are caused by basic impairments in the ability to process speech. The acquisition of two language processes—inflectional morphology and pronoun anaphora—were modelled using connectionist networks that encode grammar as statistical regularities that are learned through exposure to phonological and semantic codes.

This approach to modelling language acquisition is distinct from how it is typically conceptualized; these models did not acquire morphology and syntax through a process of learning formal grammatical constructs such as 'rules' or 'binding conditions'. Instead, they learned a more basic process having to do with mappings between the meanings and sounds of words. When put together, these skills could be described as a grammar. However this does not assume that language acquisition involves acquiring a grammar in the formal linguistics sense (Chien and Wexler, 1990). As others have argued, such a characterization seems akin to the assumption that one must learn calculus in order to learn to catch a baseball (Seidenberg, 1997). Instead, the present approach assumes that language is acquired as a series of smaller task-oriented skills such as speaking words and recognizing their meanings in sentence context.

For the same reasons, the present approach also rejects the concept that deficits in language acquisition represent deficits in grammar learning. Instead, it is suggested that processing deficits interfere with learning specific aspects of language, which has repercussions for acquiring some types of language representations. In the case of SLI, it is theorized that a phonology deficit, which is itself caused by a speech perception deficit, makes it difficult for children to acquire some grammatical regularities. With respect to past tense morphology, this deficit is thought to interfere with learning systematic relationships between meaning and sound. In the case of syntax, the phonological deficit interferes with maintaining words in working memory in order to resolve anaphoric (pronoun-referent) relationships.

In spite of their successes, the models used in this work take a simplified view of biological and psychological constraints on language acquisition. For instance, the models assume a stable brain architecture over the course of development, and in so doing do not take into account what is likely a dynamical process in which perceptual disorders change the development of brain mechanisms subserving language and memory, which in turn influence language acquisition. The networks also do not represent language learning in a situated context; in reality, language is learned in a social setting, and its use is context-conditioned. Connectionist models have tended to set aside this aspect of language, perhaps because it is felt that the critical data do not turn on such situatedness constraints. Nevertheless, I suggest that folding stronger biological and social mechanisms into existing connectionist models will only serve to further enhance their validity for studying both normal and impaired language.

Acknowledgements

This research is supported by the Canadian Language and Literacy Research Network, the Natural Sciences and Engineering Research Council of Canada and a New Investigator grant from the Canadian Institutes for Health Research. I am grateful to Mark S. Seidenberg for his collaboration on much of the research presented in this chapter.

Glossary

Anaphoric reference In syntax, the process by which a **pronoun** is used to refer to another element such as a noun (e.g, *The robber disguised himself*).

Grammar A description of the systematic properties of a language, especially as it pertains to speakers' internalized knowledge of the structure of language (known as a **mental grammar**).

Morpheme The smallest unit of language that carries a distinct meaning. Morphemes include words (e.g., *cat, break, blue*) but also grammatical markers such as prefixes that attach to the beginnings of words (*un-deniable)* and suffixes that attach to the ends of words (*silli-ness, bake-d , book-s*).

Morphology A language user's knowledge of how words and other morphemes combine to create new words. Common morphological phenomena include affixation (*talking*) and compounding (*goaltender*)

Phoneme The smallest unit of sound used to distinguish between words. For example the phonemes [t] and [m] are used to distinguish *start* and *smart.*

Phonology The knowledge of the phonemes in language, and how these phonemes can combine to create valid words. For example, this knowledge allows English speakers to distinguish between an illegal word like *ptas* and a legal word like *stap.*

Prefix A morpheme that adjoins the beginning of a word (*un-do, in-distinct*).

Pronoun A function word that is used in place of a noun, and which refers to an individual or entity either in the same syntactic context (e.g., a **reflexive** pronoun such as *herself*), or elsewhere in discourse.

Suffix A morpheme that adjoins the end of a word (*real-ity, bat-s*).

Syntax Language users' knowledge of how words are ordered to create phrases (*the nice boy*, vs. **boy the nice*) and sentences (*The car is green*, vs. **green car the is*).

Notes

1. A number of previous studies have suggested that the definition of SLI is a heterogeneous one, and that cases of developmental language disturbances cluster into several subgroups. In this paper I reserve the term SLI to refer to children whose receptive grammar deficits cannot be accounted for by low IQ, poor articulation, or frank neurological disturbance. However, exact definitions do vary across studies.
2. Asterisks are used to indicate forms that are usually judged by native speakers to be ungrammatical.

References

Berko, J (1958). The child's learning of English morphology. *Word,* 14, 150–177.

Bernstein, LE and Stark, RE (1985). Speech perception development in language-impaired children: A 4-year follow-up study. *Journal of Speech and Hearing Disorders*, 50, 21–30.

Bird, J and Bishop, DVM (1992). Perception and awareness of phonemes in phonologically impaired children. *European Journal of Disorders of Communication*, 27, 289–311.

Bishop, DVM (1997). *Uncommon understanding: Development and disorders of language comprehension in children*. Hove, East Sussex, UK: Psychology Press.

Bishop, DVM, Bright, P, James, C, Bishop, SJ, and van der Lely, HKJ (2000). Grammatical SLI: A distinct subtype of developmental language impairment? *Applied Psycholinguistics*, 21, 159–181.

Bishop, DVM, Carlyon, RP, Deeks, JM, and Bishop, SJ (1999). Auditory temporal processing impairment: neither necessary nor sufficient for causing language impairment in children. *Journal of Speech, Language and Hearing Research*, 42, 1295–1310.

Briscoe, J, Bishop, DVM, and Norbury, CF (2001). Phonological processing, language and literacy: a comparison of children with mild to moderate sensorineural hearing loss and those with specific language impairment. *Journal of Child Psychology and Psychiatry*, 42, 329–340.

Chiat, S (2001). Mapping theories of developmental language impairments: Premises, predictions and evidence. *Language and Cognitive Processes*, 16, 113–142.

Chien, Y-C and Wexler, K (1990). Children's knowledge of locality conditions in binding as evidence for the modularity of syntax and pragmatics. *Language Acquisition*, 1, 225–295.

Chomsky, N (1981). *Lectures in government and binding*. Dordrecht: Foris.

Daneman, M and Carpenter, PA (1983). Individual differences in integrating information between and within sentences. *Journal of Experimental Psychology: Learning, Memory and Cognition*, 9, 561–584.

Dick, F, Bates, E, Wulfeck, B, Utman, J, Dronkers, N, and Gernsbacher, MA (2001). Language deficits, localization, and grammar: Evidence for a distributive model of language breakdown in aphasic patients and neurologically intact individuals. *Psychological Review*, 108, 759–788.

Ellis Weismer, S (1985). Constructive comprehension abilities exhibited by language-disordered children. *Journal of Speech and Hearing Disorders*, 28, 175–184.

Evans, J (2001). An Emergent Account of Language Impairments in Children with SLI: Implications for Assessment and Intervention. *Journal of Communication Disorders*, 34 (1–2), 39–54.

Gathercole, SE and Baddeley, AD (1990). Phonological memory deficits in language disordered children: Is there a causal connection? *Journal of Memory and Language*, 29, 336–360.

Gopnik, M (1997). Language deficits and genetic factors. *Trends in Cognitive Sciences*, 1 (1), 5–9.

Gopnik, M and Goad, H (1997). What underlies inflectional errors in SLI? *Journal of Neurolinguistics*, 10, 129–137.

Hoeffner, JJ and McClelland, JL (1993). Can a perceptual processing deficit explain the impairment of inflectional morphology in development dysphasia? A computational investigation. In EV Clark (ed.), *Proceedings the 25th Annual Child Language Research Forum* (Vol. 25, pp. 38–49). Stanford, CA: Center for the Study of Language and Information.

Joanisse, MF (2000). *Connectionist Phonology.* Unpublished Ph.D. Dissertation, University of Southern California, Los Angeles CA.

Joanisse, MF (2004). Specific Language Impairments in Children: Phonology, Semantics and the English Past Tense. *Current Directions in Psychological Science*, 13(4), 156–160.

Joanisse, MF and Seidenberg, MS (1999). Impairments In Verb Morphology after Brain Injury: A Connectionist Model. *Proceedings of the National Academy of Sciences of the United States of America*, 96 (13), 7592–7597.

Joanisse, MF and Seidenberg, MS (2003). Phonology and syntax in specific language impairment: Evidence from a connectionist model. *Brain and Language*, 86(1), 40–56.

Joanisse, MF, Manis, FR, Keating, P, and Seidenberg, MS (2000). Language deficits in dyslexic children: Speech perception, phonology, and morphology. *Journal of Experimental Child Psychology*, 77 (1), 30–60.

Johnston, JR (1984). Specific language disorders in the child. In N Lass, J Northern, L McReynolds and DE Yoder (eds) *Handbook of speech-language pathology and audiology* (pp. 685–715). Philadelphia: B.C. Decker.

Jusczyk, P (1997). *The discovery of spoken language.* Cambridge, MA: MIT Press.

Kamhi, AG, Catts, HW, Mauer, D, Apel, K, and Gentry, B (1988). Phonological and spatial processing abilities in language- and reading-impaired children. *Journal of Speech and Hearing Disorders*, 53, 316–327.

King, J and Just, MA (1991). Individual differences in syntactic processing: The role of working memory. *Journal of Memory and Language*, 30, 580–602.

Leonard, LB (1982). Phonological deficits in children with developmental language impairment. *Brain and Language*, 16, 73–86.

Leonard, LB (1998). *Children with Specific Language Impairment.* Cambridge, MA: MIT Press.

Liberman, AM, Harris, KS, Hoffman, H, and Griffith, B (1957). The discrimination of speech sounds within and across phoneme boundaries. *Journal of Experimental Psychology*, 54, 358–368.

Marchman, VA, Wulfeck, B, and Ellis Weismer, S (1999). Morphological productivity in children with normal language and SLI: A study of the English past tense. *Journal of Speech, Language and Hearing Research*, 42, 206–219.

Marcus, GF, Pinker, S, Ullman, MT, Hollander, M, Rosen, TJ, and Xu, F (1992). Overregularization in Language-Acquisition. *Monographs of the Society for Research in Child Development*, 57 (4), R5–R165.

McCutchen, D, Dibble, E, and Blount, MM (1994). Phonemic effects in reading—comprehension and text memory. *Applied Cognitive Psychology*, 8 (6), 597–611.

Montgomery, JW (1995). Sentence comprehension in children with specific language impairment: The role of phonological working memory. *Journal of Speech and Hearing Research*, 38, 187–199.

Nelson, L, Kamhi, A, and Apel, K (1987). Cognitive strengths and weaknesses in language-impaired children: One more look. *Journal of Speech and Hearing Research*, 52, 36–43.

Pinker, S (1991). Rules of Language. *Science*, 253, 530–535.

Plunkett, K and Marchman, V (1993). From rote learning to system building: Acquiring verb morphology in children and connectionist nets. *Cognition*, 48, (1) 21–69.

Seidenberg, MS (1997). Language Acquisition and Use: Learning and Applying Probabilistic Constraints. *Science*, 275, 1599–1603.

Seidenberg, MS and Gonnerman, L (2000). Explaining derivational morphology as the convergence of codes. *Trends in Cognitive Sciences*, 4, 353–361.

Stark, R and Heinz, JM (1996). Perception of stop consonants in children with expressive and receptive-expressive language impairments. *Journal of Speech and Hearing Disorders*, 39, 676–686.

Sussman, JE (1993). Perception of formant transition cues to place of articulation in children with language impairments. *Journal of Speech and Hearing Research*, 36, 1286–1299.

Tallal, P, Miller, S, and Fitch, RH (1993). Neurobiological Basis of Speech—a Case for the Preeminence of Temporal Processing. *Annals of the New York Academy of Sciences*, 682, 27–47.

Thomas, MSC, Grant, J, Barham, Z, Gsödl, M, Laing, E, Lakusta, L, *et al.* (2001). Past tense formation in Williams syndrome. *Language and Cognitive Processes*, 16, 143–176.

Ullman, MT, Corkin, S, Coppola, M, Hickok, G, Growdon, JH, Koroshetz, WJ, *et al.* (1997). A neural dissociation within language: Evidence that the mental dictionary is part of declarative memory, and that grammatical rules are processed by the procedural system. *Journal of Cognitive Neuroscience*, 9 (2), 266–276.

van der Lely, HKJ (1996). Specifically language impaired and normally developing children: verbal passive vs. adjectival passive sentence interpretation. *Lingua*, 98, 243–272.

van der Lely, HKJ and Stollwerck, L (1997). Binding theory and grammatical Specific Language Impairment in children. *Cognition*, 62 (1), 245–290.

van der Lely, HKJ and Ullman, MT (2001). Past tense morphology in specifically language impaired and normally developing children. *Language and Cognitive Processes*, 16, 177–217.

van der Lely, HKJ, Rosen, S, and McClelland, A (1998). Evidence for a grammar-specific deficit in children. *Current Biology*, 8 (23), 1253–1258.

Chapter 10

A neural network model of autism: implications for theory and treatment

Ira L. Cohen

Introduction

Autism and the other Pervasive Developmental Disorders (PDD) including Childhood Disintegrative Disorder, Rett's Disorder, and PDD—Not Otherwise Specified, are characterized by a severe and, in many cases, lifelong 'triad' of impairments in socialization, communication and imagination. In autism, age of onset is, typically, between 18 and 30 months of age, although this may be an underestimate in that behavioural abnormalities may be missed at even earlier ages (Osterling and Dawson, 1994). The first behavioural disturbances noted by parents include lack of response to the child's name being called despite other evidence of apparent normal hearing; failure to anticipate being picked up; poor eye contact and lack of interest in social interaction; failure to use normal gestures such as pointing to communicate (instead pulling others to desired objects); lack of speech development or speech regression; failure to develop appropriate play skills and later development of repetitive/ritualistic behaviours such as unusual hand movements, visual fixations, rocking and resistance to change (Gillberg and Coleman, 2000). If speech is acquired, it is not used for sharing experiences and conversational purposes but, instead, is often literal, repetitive (on words, phrases or topics), echoic, emitted with an odd rhythm (prosody) and/or solely used to make requests for desired objects. As defined in the DSM-IV (American Psychiatric Association, 1994), Asperger's Disorder is a form of autism but without the developmental delay in language development or lack of interest in the environment. Such individuals are often quite intelligent but show the social interaction deficits and rituals characteristic of autism.

This unusual pattern of behaviour has been a puzzle to both theorists and practitioners alike and has led to the development of a number of theories that attempt to account for these behavioural characteristics. These theories range from the psychological to the neurological ends of the modelling spectrum. The psychological theories include those postulating theory of mind deficits (Baron-Cohen, Leslie *et al.*, 1985), executive function deficits (Ozonoff, Pennington *et al.*, 1991), stimulus overselectivity problems (Lovaas, Schreibman *et al.*, 1971), and central coherence problems (Frith and Happe, 1994). While these theories are interesting and seem to account for certain aspects of the cognitive functioning of people with autism, they often appear to be just another way of either describing the symptoms or reifying the symptoms as modular brain structures. They do not indicate how these problems would be expected to arise from specific structural differences in the brains of people with autism.

At the other end of the spectrum are neurological models of autism that hypothesize deficits in specific brain regions as the basis for autism. These include theories as to the role of the amygdala in autism (Baron-Cohen, Ring *et al.*, 2000), the hippocampus and the amygdala (Machado and Bachevalier, 2003), and right hemisphere dysfunction (Ozonoff and Miller, 1996), to name a few. While some of the functions of these brain structures are known, it is not clear how they actually perform these functions. These models therefore assume that the structures hypothesized to be involved with autism are not working properly but the mechanism by which this occurs is not specified. A lesion model could be employed and one could refer to the animal or human literature for clues as to what might happen to human behaviour when that structure is removed or destroyed. But it is not all that clear that an anomaly in the development of a brain structure is functionally the same as a lesion to that structure. Nor can we assume that a structural difference in the brain that is present at birth is functionally analogous to a lesion that is present in an adult brain (Mareschal and Thomas, 2001). It is in this regard that neural networks are interesting.

Neural networks serve as simple models of certain general principles of brain functioning, devoid of the amazing complexity of this organ. This simplicity allows us to ask certain questions about how alterations in the structural and functional constraints of the brain during development might affect the emergence of typical or of pathological behaviours. Neural network models of these constraints help establish 'conceptual bridges between neuroscience and psychological studies of behaviour' (Gluck and Myers, 2001). These models come in a vast array of architectures. There are supervised neural networks that are taught to respond by an external 'teacher' and there are unsupervised neural networks that 'self-organize' by simple repetitive exposure to an input pattern. With some neural network architectures, information always flows in one direction (e.g., feedforward networks) while in other architectures, information flows back and forth between layers (e.g., recurrent networks). Many networks distribute information throughout the system, for example parallel distributed processing (PDP) networks, while with other architectures information flows to only certain structures within the network (e.g., adaptive resonance models). The rules for changing the strength of connections between the processing elements or 'neurons' in artificial neural networks vary quite a bit. Some of these are discussed in Volume 1 and in other chapters of this volume. In addition, a number of excellent introductions to neural networks exist elsewhere (e.g., Elman, Bates *et al.*, 1997; Gluck and Myers, 2001; Levine, 1991; and Stein and Ludik, 1998).

Neural network models of autism have already begun to be developed by others (Casanova, Buxhoeveden *et al.*, 2003; Gustafsson, 1997; McClelland, 2000) and are discussed in other sections of this text. In this chapter I will review and elaborate upon a previously published neural network model of the learning characteristics of people with autism (Cohen, 1994, 1998). The model described here is a supervised PDP feedforward network and provides linkages amongst hypothesized structural difference in the brains of people with autism, their attentional and learning characteristics and their social and language deficits. It has some support from autopsy and imaging studies and also has implications for intervention. These areas of support will be discussed at the end of the chapter. Rather than focusing on particular brain regions that may be affected, this model emphasizes the problems that could result from deviations in the architecture of circuits in the brains of people with autism. By circuits, I mean groups of neurons that are connected to one another and that subserve a particular function, such as shape recognition. Indeed it would seem, perhaps, premature to speculate about the precise role of specific brain regions in autism because much of what we thought we knew about the function of parts of the brain, such as the cerebellum, is incomplete, at best (Xiang, Lin *et al.*, 2003). A more direct reason for my emphasis on circuitry, rather than brain regions, however, arises from my pondering on the mystery of why people with autism show certain types of cognitive assets and deficits.

Cognitive assets and deficits in autism

Research has shown that the cognitive abilities of people with autism are quite varied. At a gross level, most obtain sub-average global intellectual functioning on intelligence tests (about 70 per cent with IQs less than 70). Yet, not all aspects of cognition are equally affected. These individuals tend to do poorly on tests of vocabulary and comprehension but quite well on tests of rote memory and visual pattern recognition (Frith, Baron-Cohen *et al.*, 1987). Some show islets of normal to exceptional abilities (so-called savant skills) such as calendar calculating skills (Hurst and Mulhall, 1988), recall of random word strings (Hermelin and O'Connor, 1967), hyperlexia (Tirosh and Canby, 1993), computation (Kelly, Macaruso *et al.*, 1997), or drawing (Selfe, 1997). Yet, in these same individuals, social signals, such as gestures and facial expressions of various emotions, may not be adequately understood (Hobson, Ouston *et al.*, 1988; Tantam, Monaghan *et al.*, 1989).

This vast array of assets and deficits can be conceptually simplified into two categories. The types of problems that people with autism are good at solving have in common the formation of highly predictable patterns (routines, routes, computation, etc.). By contrast, the types of problems they have difficulty solving tend to have in common the need to form complex categories in which the category boundaries are noisy or 'fuzzy' (Kosko, 1993; Zadeh, 1965) and highly dependent on context. Why would a person with autism find the former easy and the latter difficult? I suggest that clues to the answer may be found by examining how neural networks learn to map input to output.

Mapping input to output: a neural network model of autism

Neural networks compute mapping functions, i.e., they learn to relate a set of data in one domain (e.g., changes in light patterns across an artificial 'retina') to another domain (e.g., internal representations of edges or corners at some intermediate layer of neurons—the hidden layer(s)—or predicted geometrical figures at the output layer). In back-propagation type networks, this is accomplished by adjusting the strengths of the many connections in the network until the output error is minimized (Werbos, 1974; Rumelhart and McClelland, 1986). Back-propagation is the generalization to multilayered networks of the 'delta rule' for adjusting network weights. In the delta rule the weights between the layers of neurons are adjusted so as to minimize the sum-squared difference between the pattern of activation that arises on the output layer and some prespecified target pattern of activation when a given pattern is presented at the input layer.

To simplify, imagine that a two-layer back-propagation network (consisting of input and output layers with one neuron in each layer and one bias neuron connected to the output neuron) is presented with data in which a linear relation exists between input (X) and output (Y): $Y = a + bX$. Using the delta rule such a network will readily learn this mathematical function simply by exposure to X and Y data. In fact, this network is functionally identical to a simple linear regression model (where Y = output neuron value; X = input neuron value; a = the connection weight between the bias neuron and the output neuron; and b = the connection weight between the input and output neurons). Now consider a simple extension of this problem. Imagine that Y can take on one of two binary states, 0 or 1, and that there are many X inputs. Further, a linear function exists that represents a boundary between two classes such that group A falls on one side of the boundary and group B falls on the other side of the boundary. Given a suitable amount of relevant data ($X_1 \ldots X_n$) describing the groups, another two-layer back-propagation network will learn to map the locations of the groups with the boundary or border separating the groups defined by a linear function that approximates the true linear function. In this case, our network is functionally similar to a linear discriminant analysis or a multiple regression model. When our prediction or classification problems become non-linear, however (e.g., $Y = a + \cos(X)$), a simple two-layer network will not solve this type of problem and a more complex model is needed, such as a three-layer network (input, hidden, and output layers). In fact, it has been shown that a single hidden layer with enough neurons can fit almost any mathematical function (Hornik, Stinchcombe *et al.*, 1989). Indeed, the more complex the function, the more neurons required in the hidden layer.

So what does all this have to do with autism and the learning assets and deficits seen in these groups of individuals? Quite a bit if we accept the following assumptions of the model:

1. *The Analogy Assumption*: The ways in which a complex system such as the brain learns to predict and to map categories is in many ways analogous to the ways in which complex artificial neural networks solve these problems. A functionalist perspective is taken here in that the similarities in how complex systems process information can be as instructive as the 'hardware' details by which they carry out such information processing.
2. *The Complexity Assumption*: From the Analogy Assumption, the ability of the brain to form appropriate categories and to predict events in the environment is, in part, dependent on its complexity, i.e., the numbers of neurons and the degree of connectivity amongst neurons in circuits that subserve a particular task, as noted above.
3. *The Normality Assumption*: Both within and between brain circuits, there exists a 'normal' range of neuron number, neural circuits, and neural connectivity that enables typical individuals to predict and to categorize events in their environments as they mature. When connectivity falls outside of this range, the prediction and categorization of events in the

child's environment has a greater chance of proceeding along a 'pathological' trajectory.

4. *The Error Surface Assumption*: In artificial neural networks, the output error of the network is a function of the strengths of the connections amongst neurons and can be visualized as an 'error surface.' This error surface can be analogized to a landscape where hills indicate poor solutions to a problem and valleys to alternate solutions of problems, some of which are better than others (Elman *et al.*, 1997). As a network learns, the strengths of the connections between neurons change and the error surface is descended until a valley or minimum error is reached. This is schematized in Figure 10.1, which shows a hypothetical plot of the output error surface for a network in which neurons A and B connect with neuron Y but not with each other. The output error from neuron Y is shown as a function of the strengths of the connections (varying from highly inhibitory to highly excitatory) between neurons A and Y and neurons B and Y. Note that the least error for the particular problem (the optimal or 'Global Minimum' error) in this figure corresponds to a situation where the connection strength from A to Y is highly excitatory and from B to Y is highly inhibitory. But there is also another solution where the connection strength from A to Y is highly inhibitory but the connection strength from B to Y is almost non-existent. This 'Local Minimum' may correspond to a solution that is not quite accurate for the problem but may actually be better in terms of the ability of the network to generalize the solution to other similar problems. It should also be noted in this figure that the initial settings for the weights that connect the neurons would affect the path that the network will take in arriving at a solution. A network with an initial highly excitatory connection from A to Y will readily enter the global minimum during its learning phase. Analogously, a complex system such as the brain may have many such local and global minima in a highly dimensional error surface. Elman, Bates *et al.* (1997) have hypothesized that the brain has evolved to avoid local minima that do not provide optimal solutions to problems. In the present model it is hypothesized that deviations in evolutionarily normal connectivity patterns affect this error surface and increase the likelihood that the ability of the child to form solutions to problems he/she will encounter will be deleteriously affected.

Following from the above points, it is proposed that people with autism have unusual differences in the numbers of neurons and/or neural connections that comprise various circuits and/or differences in the numbers of neural circuits that are formed or removed in their brains during the course of development.

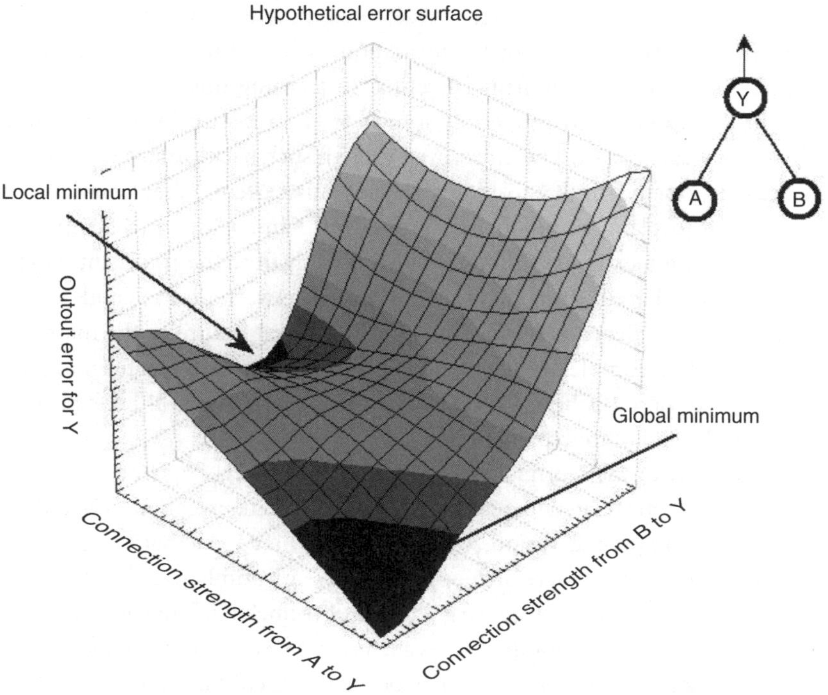

Fig. 10.1 This figure depicts a hypothetical error surface for a neural network in which two of its input neurons (A and B) and one output neuron (Y), are shown in the small inset. The error surface has both hills and valleys. The two valleys depict local, and global minima in the amount of error from neuron Y as the input weights from A and B covary. Error is smallest in the global minimum. Error is quite high when the strength of the connection between B and Y is the most excitatory, irrespective of the strength of the connection from A to Y. Note that the local and global minima are located at different positions on the error surface. Accordingly, the initial settings for the strengths of the connections between A and Y and between B and Y will influence which solution is likely to be found by the network as it learns.

They may have too few, too many, or a combination of the two depending on the brain circuits that may be affected. It is these problems with connectivity, in interaction with cues provided by the external environment, which are, mostly, responsible for their learning assets and deficits and for some of their overt behaviours. That is, these differences, embedded within a typical or 'normal' experiential environment increase the likelihood that a child will evidence the behavioural phenotype we call autism.

This proposed model of autism depends critically on the validity of the above assumptions. The Analogy Assumption has been accepted and seemingly validated by a number of theorists (Churchland and Sejnowski, 1992).

For example, Lehky and Sejnowski (1988, 1990) have been able to use a back-propagation network to extract information on curvature of objects from shaded image input to an artificial retina. In learning this task, the hidden layer neurons developed properties similar to the receptive fields of simple cells in the visual cortex. Therefore, even though such back-propagation models are not exactly similar to biological nervous systems, they mimic, and may help to explain, some of their properties (Churchland and Sejnowski, 1992). The Complexity Assumption follows directly from the arguments noted above, namely that mapping functional relations between input to and output from neural nets depends on the linear or non-linear nature of the mapping function and its complexity. The Error Surface Assumption has been argued in detail elsewhere (Cohen, 1994), is a complex issue, and will be explained further at the end of this chapter.

With respect to the Normality Assumption, the model presented here is a developmental one related to quantitative variation in neuron numbers and/or connections comprising various brain circuits. The connectivity constraints on which I am focusing here involve both local (e.g., numbers of neurons in a circuit) and global architectures (e.g., degree of connectivity between circuits) as well as timing or chronotopic issues such as synaptic pruning (Elman *et al.*, 1997). It is argued that anomalies may occur in which there exist either 'too few' or 'too many' such neurons or neural connections at various stages in a child's development. Now this concept of a 'normal' range of connectivity has been rightly criticized because it is not well defined (Mareschal and Thomas, 2001) so I will attempt to correct this error. The concept of 'too few' or 'too many' herein refers to numbers of neurons or neural connections comprising various brain circuits or to numbers of neural circuits that, at certain stages of development, are at the extreme ends of the distributions that describe those present in typically developing people: numbers that were shaped, in part, by our evolutionary history. That is, 'normal' numbers of connections are within a range that enables the typical child to form appropriate associations and concepts when exposed to the learning environment typical of human society. But how would numbers of neurons or neural connections deviate from normalcy in the brain of a child with autism?

The answer is that multiple possibilities exist. Imaging and neuropathological observations in autism have suggested possible deficits in neuronal migration during fetal development (Courchesne *et al.*, 1988; Piven *et al.*, 1990), curtailment of normal neuronal growth (Bauman, 1991), and a defect in the balance between inhibition and excitation amongst neurons (Casanova *et al.*, 2003). Another important mechanism involves problems with pruning of previously established synaptic connections.

The typically developing brain produces more neurons than it needs (Rakic, 1991) and excess neurons are removed over time through a 'pruning' process. It has been reported that newborn monkeys lose over 70 per cent of their axons in the hippocampal commissure (LaMantia and Rakic, 1990). Over-production of synaptic connections has also been noted in the neocortex of non-human primates (Rakic *et al.*, 1986). Huttenlocher has observed developmental increases and decreases in synaptic density in the human cerebral cortex (Huttenlocher, 1990; Huttenlocher and Dabholkar, 2003). While cell or connection death may be genetically programmed, the death of certain connections may be a Darwinian competitive process in which, as a function of experience, connections that have been strengthened through learning or exposure survive while those that are weak, die (Rakic, 1991). Therefore, if not already affected by pathological aberrations such as migration errors, the brains of children with autism could have too many or too few neurons and/or neural connections in various brain circuits because they are impaired in their ability to develop or strengthen new connections and/or to prune out weak or interfering connections in critical brain regions (Rapin, 1993). Figure 10.2 depicts these hypothetical problems with insufficient growth or pruning.

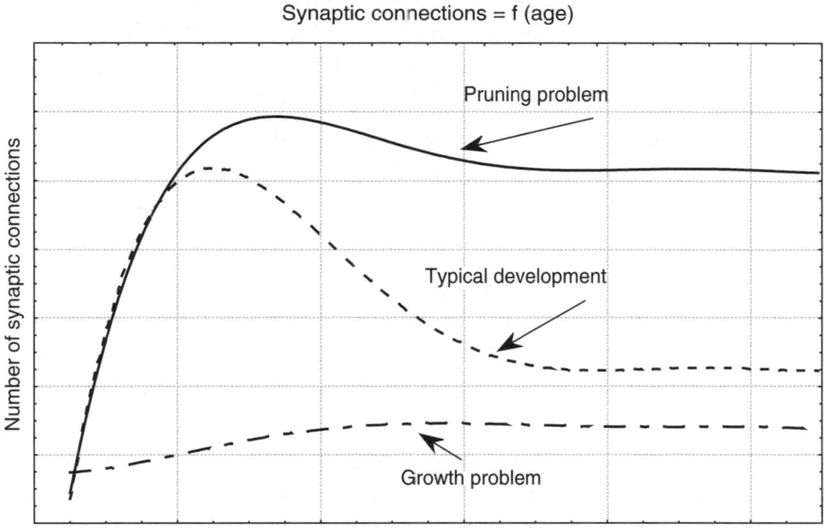

Fig. 10.2 This figure shows the hypothetical development of various neural circuits in the brains of people with autism or those with more typical development. The dashed line shows typical development. The solid line shows excessive growth of synaptic connections with delayed pruning of connections. The bottom dotted and dashed line shows insufficient growth of synaptic connections.

If we assume that growth and/or pruning problems form the basis for connectivity that deviates from normal, how would this affect behaviour and learning in the child with autism? The answer depends on the type of problems to which the child is exposed in his environment and, critically, to the quality, relevance, and amount of experiences from which the developing child's brain must form associations and concepts.

Classification and prediction problems and neural network structure

The problem of irrelevant cues

For the sake of simplicity, assume that in order to function normally most of the associations that the developing child's brain must solve during development are of two types: linear and non-linear associations. These may be further subdivided into problems of prediction (if A then B) and classification (A and B form different groups). By definition, linear classification problems have linear boundaries, i.e., groups A and B can be separated by a line, plane or hyperplane depending on the numbers of properties of A. With linear prediction problems, B is a linear function of A. When neural networks are exposed to linear problems, they are usually easily solved (given relevant input data) and do not require, nor are they helped by, hidden layers in a neural network. With back-propagation type networks, non-linear problems, as noted above, require a 'hidden layer' between input and output. The more neurons in the hidden layer, the greater the likelihood that the network will compute a non-linear function that fits the relationship between input and output.

Both types of problems are best solved when (a) the various inputs consist of both relevant and 'clean' data with very little random variation; and (b) the input data cover the range that is truly representative of the real world. But real life is often not so simple and problems can result when the amount of input data is insufficient for the model to compute true boundaries and, at the same time, there exist inputs that are actually irrelevant when applied to real world data but are spuriously correlated with the classification categories. These are the classical statistical problems of having too many predictors and not enough samples, and 'confounding', respectively. In such a case, a neural network will run into a problem. Since the network does not 'know', a priori, which inputs are relevant and which are not, it will use some of the irrelevant inputs (those spuriously correlated with the output) along with some of the relevant inputs to provide a solution.[1] When the trained network is then asked to predict real world data, however, it will fall short because the spurious inputs are not present. Therefore, we now have two problems. First, the network uses

Table 10.1 Truth table for the answer to the question 'Is Jane (or some other name) your name?'*

P (name in question)	Q (Child's response)	P if Q
Jane	Yes	True
Jane	No	False
Not Jane	Yes	False
Not Jane	No	True

*See text.

irrelevant data to predict a problem. Second, it uses only a restricted range of relevant input data. It is argued that a similar phenomenon happens to the child with autism when faced with real problems that he has to learn in a typical environment.

In working with young children with autism, it is not unusual for teachers to present their pupils with discrimination problems that, on the surface, should be easy to both learn and generalize but that cause difficulties for the child. Consider the problem of teaching such a child to answer a basic question: 'Is Jane your name?' This problem has four permutations and maps as a biconditional problem as shown in the truth table depicted in Table 10.1. That is, the child's response to the question is true if and only if the child's name is Jane and the response is 'yes' OR if the child's name is Jane, the name that is asked is not Jane, and the child answers 'no'. A back-propagation neural network with one neuron in its hidden layer easily solves this relatively simple non-linear problem. But, to a child who has not acquired language, this task is not so simple. Assuming the child's attending skills and hearing ability are not at issue, she must understand at the perceptual level that the sounds emitted by the teacher are in fact four separate words and not one continuous stream of sound. The teacher may try to help the child in computing these boundaries by speaking slowly, emphasizing each separate word and placing special emphasis on the key pronouns, much as parents speak to their infants. However, the problem is now no longer a simple non-linear one. It only becomes such a simple problem after the child's word-recognition system can parse words out from the continuous auditory stream, a task requiring many neurons and many interacting neural circuits. When the latter is accomplished, the problem may be solved. Even when the child learns the problem, however, it would be predicted that the child who possesses too many connections in her speech processing system would have difficulty with generalization and with acquisition of symbols. Why?

Notice that I indicated that the key words were highly correlated with other cues such as the loudness of the voice of the teacher and his vocal intonation. If the child is presented only with these sentences spoken by the same teacher all of the time (a sparse training set), she may learn to associate the vocal characteristics of her teacher with the correct answer or only parts of the correct answer because circuits in her brain are utilizing a limited, and sometimes irrelevant, range of stimuli in the associations that they acquire. Therefore, when these same words are spoken by another teacher or parent whose voice has a different pitch and loudness profile, the likelihood that this child would correctly respond would be markedly decreased because, to this child, the words spoken by others are in fact not precisely the same as those spoken by the teacher.[2] The reason why this literalness and generalization decrement occurs is because the child's brain structure is too complex for the data set. Much like the artificial neural network described above, irrelevant but correlated cues become part of the learned stimulus configuration. When these irrelevant cues are no longer present, a generalization decrement occurs. Difficulty in generalization would impair the ability of the child to associate words with real objects in her environment. Consequently, word meaning would be difficult to learn for this child without direct and explicit training that includes exposure to various people uttering the same words in many different situations (i.e., expanding the training set) forcing her to disregard cues that are irrelevant. The typical child does not have as much of a problem because there are fewer numbers of connections in the child's brain relative to the data set size (i.e., the experience of the child) and, hence, there is less of a tendency to form irrelevant associations.

Consider the following simple simulation of an artificial neural network required to differentiate between a 'happy' and a 'sad' face (Cohen, 1998)—see Box 10.1 for details of this simulation. Inputs were either relevant or irrelevant to the solution. The training set consisted of two numbers that described the orientation of the 'mouth' or 'eyebrows'. These relevant numbers varied from −1.0 to +1.0 for mouth orientation with negative numbers corresponding to the corners of the mouth downward and from 0 to −1.0 for eyebrow orientation with negative numbers indicating that the parts of the eyebrows closest to the temples were lower than the part closest to the nose (see Figure 10.3). Thus, the mouth input was more 'salient' in that it had greater variation in range then did the less salient, and redundant, eyebrow input. In addition to these salient, but redundant cues the first network was exposed (in the training set) to one irrelevant input (consisting of random numbers); the second network to three irrelevant inputs; the third network to five irrelevant inputs; and the fourth network to 10 irrelevant inputs. During training, the

Box 10.1 Details of neural network simulation

The models presented here consist of three-layer feedforward networks employing a back-propagation training algorithm for weight change (NeuralWare, 1993). Learning rate was set to 0.3 and momentum to 0.4. The input layer consisted of two relevant inputs corresponding to the orientation of the mouth or eyebrows and either 1, 3, 5 or 10 irrelevant inputs, depending on the architecture of the network. The relevant numbers varied from −1.0 to +1.0 for mouth orientation with negative numbers corresponding to the corners of the mouth downward and from 0 to −1.0 for eyebrow orientation with negative numbers indicating that the parts of the eyebrows closest to the temples were lower than the part closest to the nose (see Figure 10.3). The irrelevant inputs consisted of random numbers (from 0 to 1.0) derived from a uniform random numbers distribution. The number of neurons in the middle or hidden layer was varied as follows: 1, 2, 4 and 8. This variation had only a minor effect on learning because the problem was a linear one. The number of neurons in the output layer was two, corresponding to the presence of a happy or sad face. The transfer function at the hidden and output layers was a hyperbolic tangent. In keeping with typical classroom-type situations, a small data set was used. Sixteen data points were used in the training set and eighteen in the test set. Nets were run for 100 trials and stopped every ten trials to record classification accuracy. Four additional replications were run for each permutation by initializing a net to a different set of starting connection weights so that each would converge on the solution from a different starting point in weight space.

networks were periodically tested to see how well they generalized to a test set consisting of different numbers for the relevant and irrelevant inputs than were present in the training set.

Interestingly, the number of extraneous inputs affected learning speed such that, up to a point, the greater the number of extraneous inputs, the faster learning took place, as shown in Figure 10.4. Why did this happen? Although the extraneous inputs were random numbers, some were spuriously linearly correlated with the output. Random inputs 1, 3, 5 and 7 had Pearson correlations of 0.47, .39, –0.47, and –0.39, respectively, with the output. This is somewhat analogous to the teacher who provides, on some trials, cues to the correct answer by pointing to the relevant stimulus. While this speeds up learning, it can cause problems with generalization. Indeed, this was the case.

Fig. 10.3 Cartoon happy and sad faces are shown in which mouth and eyebrow orientation are manipulated to provide input to a neural network. In this example, the happy face on the left has a mouth input value of +1.0 and an eyebrow input value of 0.0. The sad face on the right has a mouth input value of −1.0 and an eyebrow input value of −1.0 (see text).

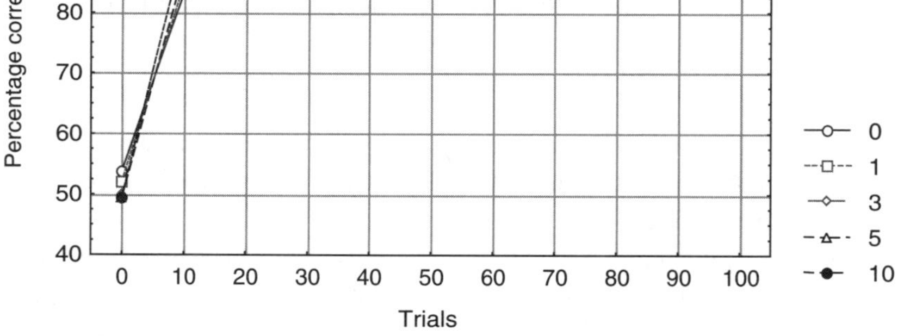

Fig. 10.4 Acquisition rate is shown as a function of number of training trials and the number of extraneous or irrelevant inputs (see text). Reprinted by permission of the publisher from Cohen, IL (1998). Neural network analysis of learning in autism. In DJ Stein and J Ludik (eds) *Neural networks and psychopathology.* New York: Cambridge University Press. Copyright 1998 by Cambridge University Press.

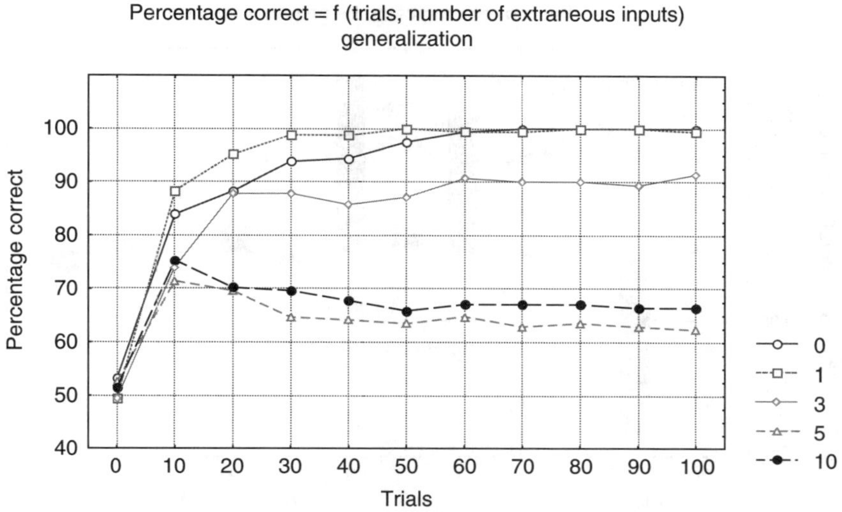

Fig. 10.5 Generalization rate is shown as a function of number of training trials and number of extraneous inputs (see text). Reprinted by permission of the publisher from Cohen, IL (1998). Neural network analysis of learning in autism. In DJ Stein and J Ludik (eds) *Neural networks and psychopathology.* New York: Cambridge University Press. Copyright 1998 by Cambridge University Press.

The number of extraneous inputs had a marked effect on both the rate and pattern of generalization during the training process. As shown in Figure 10.5, one irrelevant input had no statistically significant effect on generalization. Three extra inputs led to a lower asymptote. Five or more extraneous inputs had minimal effects on generalization, initially, but then had a marked deleterious effect as training proceeded. These data suggest that the networks utilized the spurious inputs in the training set as learning proceeded and that reliance on such data had a deleterious effect on generalization. Indeed, this was evident when I examined the effect of systematically deleting successive neurons in the input layer on generalization.

Figure 10.6 shows that the networks relied on the more salient input of mouth orientation to learn in that elimination of this input had marked deleterious effects on generalization while elimination of the eyebrow input had no effect on generalization accuracy. Elimination of successive irrelevant input neurons caused, in most instances, marked improvements in generalization accuracy, with numbers five and seven (corresponding to random inputs three and five, respectively) having the most influence—two of the four extraneous inputs that showed a moderate linear correlation with the training set outputs.

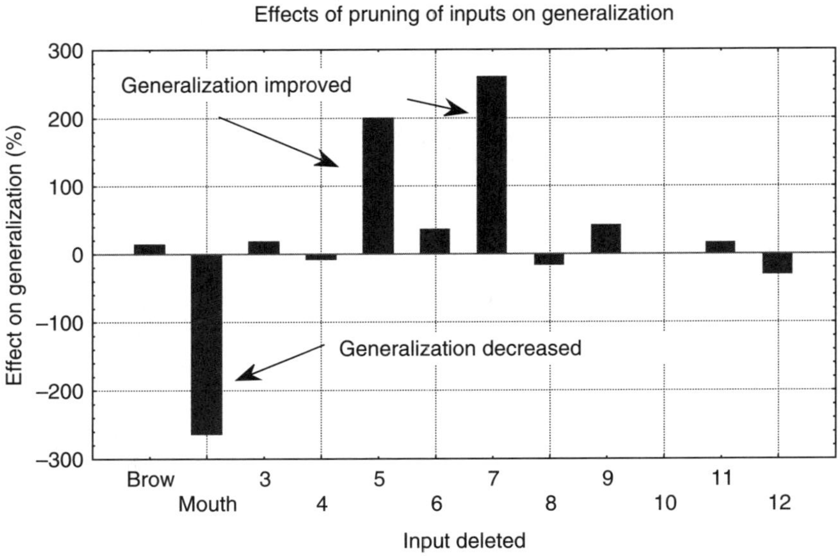

Fig. 10.6 Effects of deletion of successive inputs on generalization are shown in this figure. Positive numbers indicate percentage of improvement over baseline and negative numbers indicate percentage of decrement below baseline (see text).

It has been speculated by many (e.g., Rakic, 1991) that pruning of connections is important in 'sculpting' appropriate connections in the brain. This effect of improving generalization by pruning out irrelevant connections, however, is really quite remarkable for it indicates that knowledge regarding the correct output already exists within the pattern of connections in the network but is masked by irrelevant connections. If our network could talk would it report an 'ah hah' experience after this brain surgery? In generalizing to autism, could it be that true knowledge about the world exists in the brains of some affected people but is being blocked by irrelevant connections? Does pruning of these irrelevant connections explain the dramatic improvements seen in some children with autism when they are exposed to early and intensive behavioural intervention (Lovaas, 1987; McEachin, Smith *et al.*, 1993; Perry, Cohen *et al.*, 1995)?

The problem of fuzzy input

There are additional deleterious consequences of having a brain that is too complex for the child's experiences. Most experiences that children must process in everyday life are not easily classified into linear and non-linear problems. Most problems are not only non-linear, but they are noisy and 'fuzzy' (Zadeh, 1965; Kosko, 1993) and depend on context. For example, at what point does laughter become hysterical fear instead of delight? At what point does a white lie become just a lie? How much eye contact or information or body contact is considered 'too much' by others? Such noisy or fuzzy problems are precisely the types of problems that people with autism have difficulty acquiring, irrespective of degree of intellectual functioning. Why? Again, analysis of the anatomy of back-propagation models may provide a clue.

To repeat, back-propagation networks employ a hidden layer (or layers) that can form higher order representations of the input pattern. In forming these representations or boundary regions, the hidden layer performs a type of curve fitting, i.e., it finds a curve that fits the data points (or separates out groups in a classification problem) in the set on which it is trained (Churchland and Sejnowski, 1992). As with any other curve-fitting procedure, one desires the simplest function that will best describe the data. Why? Because, as I (Cohen, 1994), and many others have previously noted, a complex curvilinear function may fit a training set too well. This is because any data set always has some variation associated with it—variation that is not relevant to the solution of the task and that is unlikely to be present in another data sample (the 'test set'). When a complex function is used to describe this training set it describes the set so perfectly that it is relatively poorer at generalizing to a test set than would a simpler function that doesn't take into account all of the unsystematic variance in the training set. In a back-propagation network, this curve-fitting problem translates into determining the number of neurons needed in the hidden layer, as well as the number of hidden layers to use, in order to (a) best fit the data set and, at the same time, (b) best describe the test set. As the number of neurons and connections increase, the fit to the test set may deteriorate (Hertz, Krogh *et al.*, 1991). Thus, while too few neurons may not be able to solve a problem to begin with, having too many may impair generalization. This problem in overfitting the data set is also a function of the number of training trials (Hertz *et al.*, 1991) because, as the network learns, it modifies connection weights so as to minimize error on the training set (error reduction, after all, is its 'motivation'). To the extent that the test set does not precisely resemble the training set and to the extent that the data are noisy, generalization will worsen as training progresses.

In other words, variable, heavily context dependent or fuzzy cues present a similar kind of problem as irrelevant cues do to a network that has too many connections relative to the range of stimuli to which it is exposed during development in a typical environment. Both are instances of a failure to 'see the forest for the trees' although, more precisely, we should say a 'failure to see the forest for the trees and the birds and the light pattern falling on the trees and the path taken into the forest, etc.' With fuzzy or heavily context dependent situations, the overly complex network takes into account too much of the information instead of recognizing more general features in these situations. As a result, meaning becomes tied 'too much' to context, much of it irrelevant for generalization purposes. With more and more instances of exposure to such situations, 'cross-talk' (pattern interference) problems may develop, with subsequent degradation of information. If the brains of children with autism are likewise overly complex for the information present in their environment and suffer from similar problems with excessive context dependency and cross-talk, these association problems could cause some children to become puzzled and frustrated (or depressed) by their efforts at trying to fathom the rules behind such fuzzy problems. Others may be encouraged (depending on arousal state, 'temperament' and prior history) to try harder by memorizing ever more details.[3]

It has been cogently argued elsewhere that disordered brain development produces different outcomes than the same type of disorder in a fully mature brain (Mareschal and Thomas, 2001; Thomas and Karmiloff-Smith, 2002). Many models of autism and interpretations of autopsy results are, however, based on such lesion paradigms (Bachevalier, 1996) and have not considered this alternative possibility (Bauman and Kemper, 1985; Kemper and Bauman, 1998; Baron-Cohen, Ring *et al.*, 2000). In the present model, the presence of too many connections during development should produce different sequelae than those that result when circumscribed lesions occur in an otherwise normally developing or fully developed brain (Cohen, 1994, 1998). One of these sequelae could be the emergence of savant skills.

Savant skills in autism and connectivity

We have discussed how a brain structure that is too complex for certain experiences could cause the learning deficits present in children with autism. But what about their assets such as the savant skills described above? The skills evident in savants require complex pattern recognition abilities. Recall that with back-propagation networks, complex but well-defined functions can be described quite well with three layer models. To reiterate, the more complex the pattern to be recognized, the more neurons are required in the hidden layer (Hornik *et al.*, 1989). By extrapolation, development that results in an excessive number of neurons and/or neural connections in a given circuit should result in enhanced ability to define and map complex patterns. It is hypothesized, therefore, that this is the basis for the above average to savant-like skills shown by some individuals when it comes to learning and memorizing well-defined, non-linear patterns such as routes to various places or musical selections.

However, presenting such a savant with new problems to learn that are similar to, but different from, his well-learned skills require his brain to learn to map new contours. If too many neurons and neural connections are present, the possibility of cross-talk problems developing arises again with degradation of previously acquired memories as more and more new, but similar, problems are introduced. This could be the basis for the need for some savants to restrict the themes of their reproductions (Hou *et al.*, 2000), as well as for the more general forms of intolerance for changes in schedules or routines seen in people with autism.

Aberrant connectivity and the triad of impairments in autism

As noted in the beginning of this chapter, autism is characterized by deficits in communication, socialization, and imagination. The hypothetical connectivity problems discussed above have implications for the performance of each of these related skills. Here, I will present, as an example, their implications for higher-order forms of communication and for socialization and leave the implications for the remainder of the triad to the reader's imagination.

Consider the developmental consequences of forming too many neurons or neural connections in circuits of the brain important to language processing relative to the range of visual and auditory experiences afforded to most infants. Word meaning, for example, would be highly constrained and tied to irrelevant or noisy cues, as well as to a restricted range of relevant cues. Attention to a restricted range of cues could account for some echolalic verbalizations that involve literal repetitions of parts of sentences said by others. Echolalia in such cases is an indication that the child understands that he is supposed to emit a vocalization from his lips when presented with a request but has not developed a higher-order symbol system. Lack of development of a symbol system is, then, secondary to a failure to generalize because of lack of attention to all relevant cues in a sentence and to the acquisition of spurious associations. It also could lead to more and more problems with comprehension and a subsequent loss of interest in communicating, a regression phenomenon often reported by parents of children with autism.

An example of how connectivity would impact on an important social skill, the ability to recognize emotions, is shown in Figures 10.7 and 10.8 and is an elaboration on the scenario described above. These figures depict a situation where happy or sad facial expressions are quantified by two arbitrary multivariate inputs, e.g., activation of two different facial muscle groups in

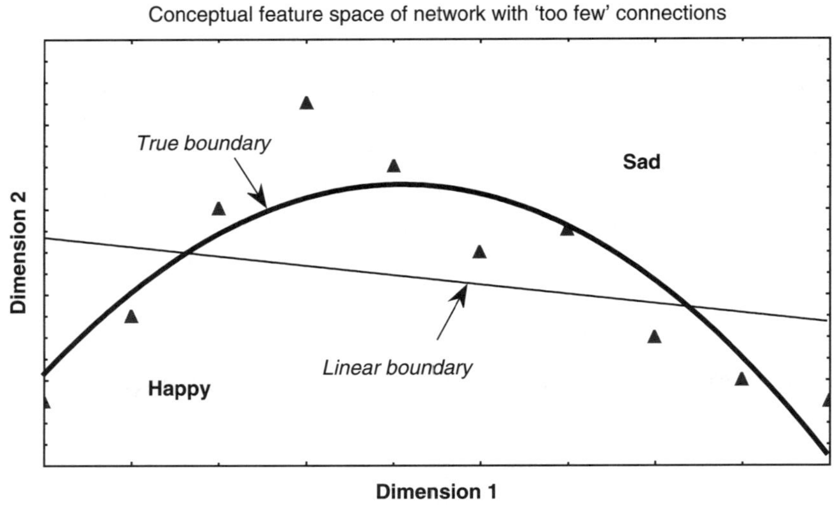

Fig. 10.7 A hypothetical two-dimensional weight space for the problem of differentiating between angry and sad faces with a noisy or fuzzy boundary (closed triangles). The quadratic function indicates the general trend in the discrimination between these two states for typical people (the true boundary). The child with two few neurons in emotion recognition brain circuits lacks the computational power to identify this curvilinear boundary (represented by the linear boundary) and so has difficulty discriminating these states (see text). Reprinted by permission of the publisher from Cohen, IL (1998). Neural network analysis of learning in autism. In DJ Stein and J Ludik (eds) *Neural networks and psychopathology.* New York: Cambridge University Press. Copyright 1998 by Cambridge University Press.

association with voice tone. The closed triangles demarcate a noisy region that separates these two types of emotional expressions. The noisy variation may be attributed to differences, say, in the magnitude of the movement of the muscles in the faces of different people and/or their tone of voice. In the typically developing child, emotion recognition circuits possess a normal number of neurons and neural connections and these circuits solve this discrimination problem by establishing a non-linear boundary that recognizes a quadratic trend in the distinction between these two emotional states, ignoring the subtle variation. The child with too few neurons in his emotion recognition circuits (Figure 10.7) doesn't possess enough computational power to adequately differentiate these states and so the boundary established by these circuits is linear with substantial overlap amongst the two true states. The child with too many neurons in these circuits (Figure 10.8) has the opposite problem. His emotional circuits take into account all of the noisy aspects of the border between these two states, as illustrated by the spline curve. As a result, this child learns to attend to too

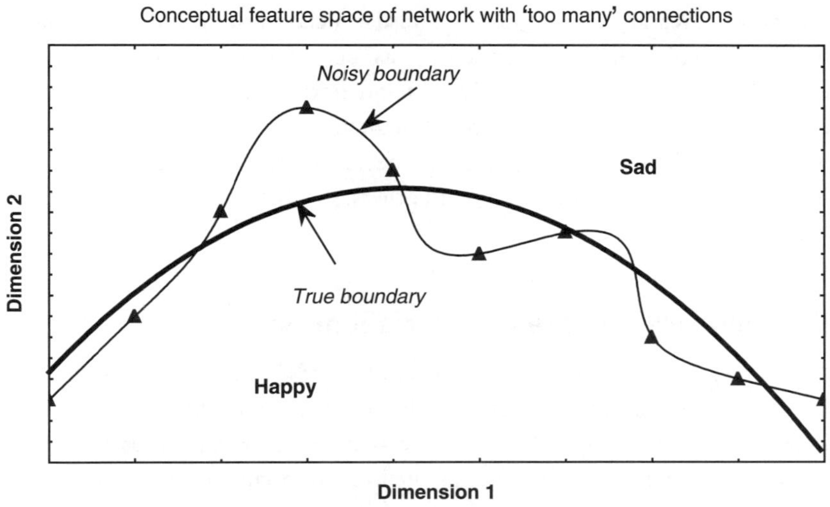

Fig. 10.8 A hypothetical two-dimensional weight space for the problem of differentiating between angry and sad faces with a noisy or fuzzy boundary (closed triangles). The quadratic function indicates the general trend in the discrimination between these two states for typical people (the true boundary). Emotion recognition circuits that have too many neurons attend to all of the noise in this curvilinear boundary (represented by the spline function noisy boundary) and so have difficulty generalizing these states (see text). Reprinted by permission of the publisher from Cohen, IL (1998). Neural network analysis of learning in autism. In DJ Stein and J Ludik (eds) *Neural networks and psychopathology.* New York: Cambridge University Press. Copyright 1998 by Cambridge University Press.

many details and, so, generalization of these affective states is hampered. In either situation, knowledge concerning affective associations is idiosyncratic and fails to generalize well. Therefore, other normally developed neural circuits that rely on this information to form symbols may show poor performance because they lack relevant information concerning the affective salience of such significant stimuli in the child's environment, an 'embrainment' issue (Churchland and Sejnowski, 1992; Eichenbaum, 1993).

Too few neurons and neuronal connections therefore set an upper limit on the number and quality of affective associations that can be acquired by a circuit. If too many neurons and neuronal connections exist in certain brain circuits, then the number and quality of affective associations that can be learned and stored may be limited, associations could be highly idiosyncratic, and there may also be deterioration in the quality of these associations with time, depending on the memory storage capacity of the circuit (Carpenter and Grossberg, 1991). This deficiency in processing of affective responses from others could readily account for the unusual social learning and performance deficits shown by these children, such as lack of awareness as to when to establish eye contact, how closely to stand next to others, etc. In a causal chain, failure to comprehend the function of fuzzy social cues, and to associate *affective meaning* with them, would be expected to lead to failure of the child to have a desire to share experiences with others, so-called joint attention deficits (Mundy, Sigman *et al.*, 1990).

Error surface and the broader phenotype

At this point I return to the fourth assumption of my model. That is, that the developmental path that results from deviations in connectivity will depend, in part, on the shape of the 'error surface' that describes learning as a function of the connection strengths amongst neurons in a circuit (see Figure 10.1). Simple circuits with relatively few neurons have simple error surfaces; more complex circuits have more complex error surfaces (Kordos and Duch, 2004).

Throughout this chapter I have argued that too many connections are pathological in that they produce deleterious effects on learning. But what if this were the case in only some instances? What if, in the majority of cases, having too many connections has an adaptive advantage for individuals but in a small percentage of cases, certain initial connection strengths between neurons lead to a drastically different 'autistic' path in behavioural development with a child becoming 'stuck' in a local or global minimum that impairs generalization for fuzzy, context-dependent problems? After all, the cortex is much more developed in humans than in other primates so a brain that has increasingly higher numbers and a greater complexity of connections is, in some sense, more 'intelligent' and better able to cope with and manipulate its environment to its advantage than a brain with less complexity. Indeed, in simulations I have described elsewhere (Cohen, 1998), some networks that had large numbers of neurons in their hidden layers relative to the training set size provided relatively good generalization to a test set. These nets clearly took a different pathway in the network's 'error surface' landscape than did networks that did not generalize as well. Up to a point, I found that the larger the number of neurons in the hidden layer, the greater the variability in generalization to the test set. Beyond a certain point, however, variability decreased, despite the increase in complexity of the network.

Autism is associated with macrocephaly that is also familial (Fidler *et al.*, 2000). It is likewise associated with enlarged brain size (Courchesne, Carper *et al.*, 2003). MRI studies of the human brain indicate that brain size is highly heritable (Pfefferbaum, Sullivan *et al.*, 2000) and positively correlated with IQ (Pennington, Filipek *et al.*, 2000). Across primate species, increases in brain volume are associated with increases in intrahemispheric connectivity (Rilling and Insel, 1999). Thus, if we assume that large and complex brains are, in part, familial, then behavioural outcomes would be expected to be quite varied in these 'big-brained' individuals both within and across generations of family members. This could result in good outcomes (e.g., people who may be unusually gifted in their ability to handle complex non-linear problems such as mathematics or computer science), less advantageous ones (e.g., autism and/or mental retardation) or a combination of autism or autistic-like behaviour and giftedness (e.g., many typical Asperger's cases), depending on the number and types of 'minima' in their error surfaces. This scenario could account for the 'broader phenotype' of social and cognitive impairment sometimes observed in relatives of the child with autism (Folstein and Rutter, 1988).

Predictions of the model

Thus, quantitative variation in one basic parameter, number of neurons and neural connections in a neural circuit, can result in complex effects on learning and generalization. These complex outcomes mimic several well-validated characteristics of the learning style of children with autism:

1. Greater attention to idiosyncratic, than to socially relevant, stimuli;
2. Stimulus overselectivity or a lack of drive for central coherence;
3. Problems in acquiring fuzzy concepts;
4. Development of savant skills;
5. Problems with generalization of previously acquired skills;
6. Rigidity and resistance to change;
7. Social and communication deficits; and
8. Difficulty in learning complex higher-order concepts.

At this point, I will make some predictions that follow logically from the model and from the simulations I have conducted. Some of these predictions have been described elsewhere (Cohen, 1998) or in this chapter and I will summarize them here.

1. The brains of people with autism should be larger than average, especially at earlier ages. This is depicted in Figure 10.2. Since I talk about circuits, the increases should occur in both grey and white matter. There is evidence to support this claim (Aylward, Minshew *et al.*, 2002; Courchesne, Carper *et al.*, 2003).

2. In general, brain size during early development should be positively correlated with aspects of cognitive functioning associated with autism. There also exist data that support this assumption in children with autism (Deutsch and Joseph, 2003). It is also predicted that brain size should be associated with the kinds of information processing issues discussed in this paper in people with autism (i.e., restricted range of attention and attention to irrelevant details), their non-autistic relatives, or in the general population. This has not been examined.
3. Since there are many ways in which brain growth and pruning could result in the problems discussed in this chapter, autism likely results from complex interactions amongst a variety of different genes that affect these processes, rather than a single small subset of genes. Some of these genes have been implicated in some cases of autism or disorders associated with autism and include HOX-A1 (Rodier, 2000), REELIN (Fatemi *et al.*, 2001), MAO-A (Cohen *et al.*, 2003), FMR-1, the gene associated with fragile X syndrome (Churchill *et al.*, 2002; Irwin *et al.*, 2002), and MeCP2, the gene associated with Rett syndrome (Johnston *et al.*, 2001). A variety of other genes have also been implicated in growth and pruning but have not yet been examined for their association with autism, such as the ephrin family of tyrosine kinase receptors implicated in the outgrowth and pruning of axons (Gao, Yue *et al.*, 1999). Teratological agents that affect these same developmental processes have also been implicated (Stromland *et al.*, 1994).
4. The brains of people with autism may not be 'pathologically different' in many cases and the types of structural differences seen in some cases have adaptive advantages in some family members. If brain size is familial, and the landscape of the 'error surface' plays a role in emergent behaviour patterns, then, in very large samples of families with a proband having autism, there should be an increase in the range of those assets and deficits normally associated with autism, relative to typical families where autism is not present. That is, in such families, there should be higher than average rates of the 'broader phenotype'; exceptional abilities in solving problems involving complex non-linear associations; and poor ability to solve 'fuzzy' social comprehension problems.
5. Savant skills should correlate with the size of certain brain regions. For savants, the number of neurons and neural connections should be increased in those networks responsible for learning visual, auditory, spatial and/or temporal patterns, depending on the particular skills of the savant. In other words, those with savant visual memory skills should

have enlarged development of visual, but not haptic memory circuits.

6. The emergence of rigidity and resistance to change should, in part, be predicted by the extent to which a child with autism forms irrelevant associations.
7. Regression in language and social development in children with autism should also be predicted by the tendency for the child to acquire irrelevant associations. This tendency should be an early sign of autism.
8. Intervention for children with autism would have to be intensive (many hours of the day) and geared toward focusing their attention on forming relevant associations and on eliminating irrelevant associations. The intensity would restrict their opportunity for learning new irrelevant associations and the heavily structured teaching would encourage the formation of relevant associations. Again, there is evidence to support this concept (Lovaas, 1987; McEachin *et al.*, 1993; Smith *et al.*, 2000). Such intensive therapy is hypothesized to work by focusing the child's attention on a larger and more varied data set than the child is used to, thereby forcing him to attend to all relevant aspects of the problem space. Then, given suitable problems to solve (e.g., ones dealing with affective recognition of emotional states or learning to recognize words pronounced by different people), constant repetition of these many different patterns along with immediate feedback should lead to higher-order recognition by weakening or pruning previously established, irrelevant connections that do not 'pay-off' and maintaining or strengthening those that do. That is, it is assumed that such intensive work may stimulate a sluggish neurodevelopmental process.
9. Where neural capacity is sufficient, interventions should focus on teaching the child to attend to relevant dimensions of a stimulus and to ignore irrelevant dimensions. This means systematically varying the physical dimensions of the relevant stimuli (making them louder and softer, larger and smaller, as the case may be) and varying the teacher and setting. External guidance cues such as a finger point, a fixed tone of voice, a fixed teacher, etc. should *not* be used. This is because the external cues will rapidly become part of the learned configuration and interfere with generalization. There is evidence to support this prediction. It has been found that when external cues are presented to aid discrimination learning, (so-called extra-stimulus prompts), children with autism have marked difficulty in maintaining their task performance when such cues are eliminated (Schreibman *et al.*, 1982). Solutions to this problem have

included increasing the salience of the stimuli by exaggerating their features—also known as within-stimulus prompts (Schreibman, 1975). However, neural network simulations suggest that this type of intervention should be minimized as quickly as possible and that the ways in which the stimuli are presented should be varied in order to avoid the formation of irrelevant associations.

10. The types of cues that children with autism will be likely to attend to during exposure to tasks will depend on the relative salience of the cues, prior preferences for these cues, prior learning, and the extent to which the cues are correlated with the task or with other relevant or irrelevant stimuli that predict the task. On average, older, higher-functioning children should have a relatively superior ability to recognize details within a complex stimulus pattern because of their extensive experience in attending to both relevant and irrelevant stimuli.
11. For some children with autism, reasonably accurate knowledge of some aspects of their world is present but masked by the presence of irrelevant associations. The task of education and/or pharmacotherapy will be to selectively eliminate these faulty connections.

Alternative models

The present model provides possible explanations for many of the features of people with autism. It is not, of course, totally comprehensive. It does not, for example, explain other characteristics of autism such as why hand flapping and hand mannerisms are frequent forms of stereotyped behaviours in this population. It also does not deal directly with hyperarousal issues, a problem often seen in autism. However, excess numbers of connections could have a direct influence on arousal and this issue is covered in other similar neural network models.

As noted in the beginning of this chapter, several other computational models of autism have been proposed as alternate explanations for the behavioural phenomena that we have discussed. In Gustafsson's model (Gustafsson, 1997), the problem in the person with autism is in the inadequate self-organizational development of his 'cortical feature maps'. To represent this type of mapping, Gustafsson used a Kohonen-type neural network, in which units compete with one another to represent a given input vector. In Kohonen networks, the mechanism responsible for competition between neurons is lateral inhibition, and in Gustafsson's model such inhibition is thought to be excessive in autism. This results in networks that attend to a restricted set of features in a stimulus configuration (i.e., too much discrimination).

Thus, Gustafsson's model results in effects similar to what I have described. The difference is that Gustafsson's model emphasizes excessive inhibition as the primary culprit. Whether or not this is true is unclear but does seem at variance with what we know about autism, namely, that these people are at increased risk for developing seizures (Gillberg and Coleman, 2000; Casanova *et al.*, 2003) in which lack of inhibition seems to be a problem (Casanova *et al.*, 2003).

In a related model, Casanova *et al.* (2003) has suggested that autism may result from an imbalance between the processes of excitation and inhibition towards too little inhibition, rather than too much. His computer models suggest that this imbalance biases the system toward a focus on a restricted set of stimulus features, as in Gustafsson's model, and affects the columnar organization of the cortices of people with autism. He presents evidence, based on autopsies of cases with autism and controls (Casanova *et al.*, 2003) that cortical minicolumnar organization is disrupted in autism: too many minicolumns with fewer cells per column. This research is potentially important (more autopsy data is clearly needed) but also fits in quite well with the model presented in this chapter, i.e., autism may be a consequence of an abnormal balance between too many and too few processing elements in brain circuits.

Unfortunately, we know very little about the brains of people with autism based on autopsy information. The numbers of brains that have been systematically examined in the same way and in the same lab are really quite small, as are the numbers of relevant controls. This is a problem mainly because most researchers agree that autism is a heterogeneous disorder both phenotypically and etiologically. Further, people that come to autopsy may represent a biased sample of more severely affected individuals. Such people may be more likely to accidentally die due to lack of awareness of danger in their environments (e.g., accidental drowning) or from CNS problems associated with more severe cases of autism, e.g., seizures.

In McClelland's model (McClelland, 2000), the highly specific learning and generalization problems characteristic of children with autism are secondary to the ways in which their brains code incoming information. According to McClelland, the brains of these children tend to excessively use a 'conjunctive representation' rule; in which neurons becomes activated only when particular combinations of features co-occur. People without autism are hypothesized to do this less often and, instead, attend more to the individual elements of the stimulus array, a 'componential representation' rule. An over-focus on conjunctive coding would impair generalization because any slight change in the stimulus array would not match the original coding.

This model bears some similarities with those above but differs in one important respect. It does not indicate how the brain would be structured so as to force this type of conjunctive coding. While this type of coding problem may occur, it remains to be demonstrated what the mechanism might be.

All of these models note problems with generalization and excessive attention to detail as a starting point for modelling. The first three are, in a sense, conceptually similar with respect to causation. They all focus on the concept that connectivity constraints are responsible for attention to a restricted range of details in autism. They differ only in the types of constraints that produce these problems. Such research is sorely needed in order to better define the role that computational models may play in helping to explain and predict the vast variety of behavioural and neurodevelopmental characteristics of autism.

Conclusion

In keeping with the other chapters in this volume, I will conclude by reviewing how this model of autism fits within four basic questions:

1. What functional brain constraints operate on the process of representation development in my model? Within the levels outlined by Elman, *et al.* (1997), the current model emphasizes local, global and timing constraints. The model suggests differences from 'normality' exist in connectivity patterns in the brains of people with autism. These can occur due to differences in numbers of neurons or packing density (local constraints), connections between different circuits in the brain (global constraints), and when and how growth and pruning of connections or cell death occurs (timing issues).
2. What embodiment or situatedness constraints operate on the process of representation development in my model? Embodiment issues are addressed in this model as an embrainment issue. That is, it is speculated that lower level processing problems resulting from connectivity differences can impact on the adequacy of representation development at higher cortical levels, even when connectivity differences are not evident at these higher levels. Situatedness is addressed in terms of the nature of the problems that a child must solve during his/her development. It is suggested that, in a typical environment, social cues tend to be highly context dependent and 'fuzzy' in nature. These present a problem for brain circuits that are 'too complex' for the average social environment. Such circuits will tend to associate too many irrelevant details with relevant ones in such fuzzy social situations, leading to degradation in generalization to similar social situations that do not share all of these features in common.

By contrast, complex but lawful/predictable problems, such as those that require memorization, or that require intense effort to arrive at lawful solutions (e.g., engineering, or mathematics problems), will be readily acquired by such complex circuits.

3. How does my model implement these constraints? In the models that I have explored (Cohen, 1998), the problems of 'too few' and 'too many' connections have been implemented by varying the numbers of irrelevant input neurons and/or the number of hidden neurons. The sparseness and 'fuzziness' of the input space has been varied by presenting training data that are limited in range, relative to the test set and by varying the degree to which clouds of input data overlap one another.
4. What have we learned from the modelling exercises? These modelling exercises suggest that autism may be one of many social and language learning problems that can emerge from changes in the mechanisms responsible for development and pruning of connections in the brain. They also suggest that, in some situations, appropriate knowledge of the world exists in the patterns of connections in the brains of people with autism but is blocked by the presence of irrelevant connections.
5. Does the model make concrete predictions? The model does indeed make a number of predictions and these are enumerated above.

Acknowledgements

This research was supported by funds from the New York State Office of Mental Retardation and Developmental Disabilities. The author is most grateful to the following friends and colleagues for their helpful comments and criticisms during the preparation of this chapter: Manuel Casanova, Mary Coleman, Michael Flory, Denis Mareschal, Vicki Sudhalter and the two anonymous reviewers.

Glossary

Input/output/hidden layers Neural network architectures are divided into two or more layers with one or more processing elements or neurons per layer. The simplest network has two layers: input and output. Such networks have interesting properties but perform poorly when presented with problems that are non-linear in nature. Adding one or more layer(s) of neurons between the input and output layers, the hidden layer(s), enables networks to solve such non-linear problems. Neurons in such layers receive inputs from the input layer, and send outputs to the output layer. In some network architectures, hidden layer neurons connect with some or all other neurons in the same hidden layer. Kohonen architectures, for example, implement lateral inhibitory mechanisms wherein a neuron in a hidden layer that is excited by a certain input pattern inhibits its neighboring neuron according to a 'Mexican hat' function.

Bias connection The bias neuron in neural networks sends connections to each of the neurons in the hidden and output layers. When these connections are allowed to vary in strength during the learning process, they permit each neuron to have a different activation threshold, allowing for improved learning of various input patterns. This bias connection is directly analogous to the constant in a linear regression equation, a factor that permits a greater degree of precision in predicting output from input.

Fuzzy logic In normal Boolean logic, truth statements are binary and take on one of two values, i.e., true or false (0 or 1) Dr. Lotfi Zadeh (1965) proposed the use of the term 'fuzzy logic' to define variations or gradations in truth that people use in normal discourse ranging from completely true to completely false. With this system, one can take into account the uncertainties in normal language such as describing someone as somewhat thin, very thin, etc.

Notes

1. Not all of the relevant inputs will be used because their importance may be diminished by other relevant or irrelevant inputs.
2. This is analogous to our difficulty in understanding foreigners who have a strong accent and who speak our language but whose intonation or vocal emphasis sometimes changes the meaning of a word.
3. This dependence of learned detail on arousal has been instantiated in adaptive resonance models such as ART-1 (Carpenter and Grossberg, 1991).

References

American Psychiatric Association (1994). *Diagnostic and statistical manual of mental disorders-IV*. Washington, DC: American Psychiatric Association.

Aylward, EH, Minshew, NJ, Field, K, Sparks, BF and Singh, N (2002). Effects of age on brain volume and head circumference in autism. *Neurology*, 59, 175–183.

Bachevalier, J (1996). Brief report: medial temporal lobe and autism: a putative animal model in primates. *Journal of Autism and Developmental Disorders*, 26, 217–220.

Baron-Cohen, S, Leslie, AM, and Frith, U (1985). Does the autistic child have a 'theory of mind'? *Cognition*, 21, 37–46.

Baron-Cohen, S, Ring, HA, Bullmore, ET, Wheelwright, S, Ashwin, C and Williams, SC (2000). The amygdala theory of autism. *Neuroscience and Biobehavioral Reviews*, 24, 355–364.

Bauman, M and Kemper, TL (1985). Histoanatomic observations of the brain in early infantile autism. *Neurology*, 35, 866–874.

Bauman, M (1991). Microscopic neuroanatomic abnormalities in autism. *Pediatrics*, 87, 791–796.

Carpenter, GA and Grossberg, S (1991). *Pattern recognition by self-organizing neural networks.* Cambridge, MA: MIT Press.

Casanova, MF, Buxhoeveden, D and Gomez, J (2003). Disruption in the inhibitory architecture of the cell minicolumn: implications for autism. *The Neuroscientist*, 9, 496–507.

Casanova, MF, Buxhoeveden, D, Switala, AE and Roy, E (2002). Neuronal density and architecture (Gray Level Index) in the brains of autistic patients. *Journal of Child Neurology*, 17, 515–521.

Churchill, JD, Grossman, AW, Irwin, SA, Galvez, R, Klintsova, AY, Weiler, IJ *et al.* (2002). A converging-methods approach to fragile X syndrome. *Developmental Psychobiology*, 40, 323–338.

Churchland, PS and Sejnowski, TJ (1992). *The computational brain.* Cambridge, MA: MIT Press.

Cohen, IL (1994). An artificial neural network analogue of learning in autism. *Biological Psychiatry*, 36, 5–20.

Cohen, IL (1998). Neural network analysis of learning in autism. In DJ Stein and J Ludik (eds) *Neural networks and psychopathology*, pp. 274–315. New York: Cambridge University Press.

Cohen, IL, Liu, X, Schutz, C, White, BN, Jenkins, EC, Brown, WT *et al.* (2003). Association of autism severity with a monoamine oxidase A functional polymorphism. *Clinical Genetics*, 64, 190–197.

Courchesne, E, Yeung-Courchesne, R, Press, GA, Hesselink, JR, and Jernigan, TL (1988). Hypoplasia of cerebellar vermal lobules VI and VII in austism. *New England Journal of Medicine*, 318, 1349–1354.

Courchesne, E, Carper, R and Akshoomoff, N (2003). Evidence of brain overgrowth in the first year of life in autism. *JAMA*, 290, 337–344.

Deutsch, CK and Joseph, RM (2003). Brief report: cognitive correlates of enlarged head circumference in children with autism. *Journal of Autism and Developmental Disorders*, 33, 209–215.

Eichenbaum, H (1993). Thinking about brain cell assemblies. *Science*, 261, 993–994.

Elman, J, Bates, EA, Johnson, MH, Karmiloff-Smith, A, Parisi, D and Plunkett, K (1997). *Rethinking innateness: A connectionist perspective on development.* Cambridge, MA: MIT Press.

Fatemi, SH, Stary, JM, Halt, AR and Realmuto, GR (2001). Dysregulation of Reelin and Bcl-2 proteins in autistic cerebellum. *Journal of Autism and Developmental Disorders*, 31, 529–535.

Fidler, DJ, Bailey, JN and Smalley, SL (2000). Macrocephaly in autism and other pervasive developmental disorders. *Developmental Medicine and Child Neurology*, 42, 737–740.

Folstein, SE and Rutter, M (1988). Autism: familial aggregation and genetic implication. *Journal of Autism and Developmental Disorders*, 18, 3–30.

Frith, U, Baron-Cohen, S and Donnellan, AM (1987). Perception in autistic children. In DJ Cohen (ed.) *Handbook of autism and pervasive developmental disorders*, pp. 85–102. New York: Wiley.

Frith, U and Happe, F (1994). Autism: beyond 'theory of mind'. *Cognition*, 50, 115–132.

Gao, PP, Yue, Y, Cerretti, DP, Dreyfus, C and Zhou, R (1999). Ephrin-dependent growth and pruning of hippocampal axons. *Proceedings of the National Academy of Sciences USA*, 96, 4073–4077.

Gillberg, C and Coleman, M (2000). *The biology of the autistic syndromes*, 3rd edn. London: Mac Keith Press.

Gluck, MA and Myers, CE (2001). *Gateway to memory: An introduction to neural network modeling of the hippocampus and learning*. Cambridge, MA: MIT Press.

Gustafsson, L (1997). Inadequate cortical feature maps: a neural circuit theory of autism. *Biological Psychiatry*, 42, 1138–1147.

Hermelin, B and O'Connor, N (1967). Remembering of words by psychotic and subnormal children. *British Journal of Psychology*, 58, 213–218.

Hertz, J, Krogh, A and Palmer, RG (1991). *Introduction to the theory of neural computation*. Redwood City, CA: Addison-Wesley.

Hobson RP, Ouston J, and Lee A (1988). Emotion recognition in autism: coordinating faces and voices. *Psychological Medicine*, 18, 911–23.

Hornik K, Stinchcombe M, and White H (1989). Multilayer feedforward networks are universal approximators. *Neural Networks*, 2, 359–68.

Hou C, Miller BL, Cummings JL, Goldberg M, Mychack P, Bottino V *et al.* (2000). Autistic savants. *Neuropsychiatry Neuropsychology and Behavioral Neurology*, 13, 29–38.

Hurst LC and Mulhall DJ (1988). Another calendar savant. *British Journal of Psychiatry*, 152, 274–77.

Huttenlocher PR (1990). Morphometric study of human cerebral cortex development. *Neuropsychologia*, 28, 517–27.

Huttenlocher PR and Dabholkar AS (2003). Regional differences in synaptogenesis in human cerebral cortex. *Journal of Comparative Neurology*, 387, 167–78.

Irwin SA, Idupulapati M, Gilbert ME, Harris JB, Chakravarti AB, Rogers EJ *et al.* (2002). Dendritic spine and dendritic field characteristics of layer V pyramidal neurons in the visual cortex of fragile-X knockout mice. *American Journal of Medical Genetics*, 111, 140–46.

Johnston MV, Jeon OH, Pevsner J, Blue ME, and Naidu S (2001). Neurobiology of Rett syndrome: a genetic disorder of synapse development. *Brain Development*, 23 Suppl 1:S206–13., S206-S213.

Kelly SJ, Macaruso P, and Sokol SM (1997). Mental calculation in an autistic savant: a case study. *Journal of Clinical Experimental Neuropsychology*, 19, 172–84.

Kemper TL and Bauman M (1998). Neuropathology of infantile autism. *Journal of Neuropathology and Experimental Neurology*, 57, 645–52.

Kordos M and Duch W (2004). On some factors influencing MLP error surface. *The Seventh International Conference on Artificial Intelligence and Soft Computing (ICAISC)*. Zakopane, Poland. June. http://www.phys.uni.torun.pl .

Kosko B (1993). Fuzzy Thinking: The New Science of Fuzzy Logic. Hyperion, New York.

LaMantia AS and Rakic P (1990). Axon overproduction and elimination in the corpus callossum of the developing rhesus monkey. *Journal of Neuroscience*, 10, 2156–75.

Lehky SR and Sejnowski TJ (1988). Network model of shape from shading: Neural function arises from both receptive and projective fields. *Nature*, 333, 452–54.

Lehky SR and Sejnowski TJ (1990). Neural network model of visual cortex for determining surface curvature from images of shaded surfaces. *Proceedings of the Royal Society of London*, B 240, 251–78.

Levine DS (1991). *Introduction to Neural and Cognitive Modeling.* Lawrence Erlbaum Associates, Inc., Hillsdale, NJ.

Lovaas OI (1987). Behavioral treatment and normal educational and intellectual functioning in young autistic children. *Journal of Consulting and Clinical Psychology*, 55, 3–9.

Lovaas OI, Schreibman L, Koegel RL, and Rehm R (1971). Selective responding by autistic children to multiple sensory input. *Journal of Abnormal Psychology*, 77, 211–22.

Machado CJ and Bachevalier J (2003). Non-human primate models of childhood psychopathology: the promise and the limitations. *Journal of Child Psychology and Psychiatry*, 44, 64–87.

Mareschal D and Thomas MSC (2001). Self-organization in normal and abnormal cognitive development. In AF Kalverboer and A Gramsbergen, eds. *Handbook of Brain and Behaviour in Human Development.* pp. 743–766, Kluver Academic Press.

McClelland JL (2000). The basis of hyperspecificity in autism: a preliminary suggestion based on properties of neural nets. *Journal of Autism and Developmental Disorders*, 30, 497–502.

McEachin JJ, Smith T, and Lovaas OI (1993). Long-term outcome for children with autism who received early intensive behavioral treatment. *American Journal on Mental Retardation*, 97, 359–72.

Mundy P, Sigman M, and Kasari C (1990). A longitudinal study of joint attention and language development in autistic children. *Journal of Autism and Developmental Disorders*, 20, 115–28.

Osterling J and Dawson G (1994). Early recognition of children with autism: a study of first birthday home videotapes. *Journal of Autism and Developmental Disorders*, 24, 247–57.

Ozonoff S and Miller JN (1996). An exploration of right-hemisphere contributions to the pragmatic impairments of autism. *Brain and Language*, 52, 411–34.

Ozonoff S, Pennington BF, and Rogers SJ (1991). Executive function deficits in high-functioning autistic individuals: relationship to theory of mind. *Journal of Child Psychology and Psychiatry*, 32, 1081–105.

Pennington BF, Filipek PA, Lefly D, Chhabildas N, Kennedy DN, Simon JH *et al.* (2000). A twin MRI study of size variations in the human brain. *Journal of Cognitive Neuroscience*, 12, 223–32.

Perry R, Cohen IL, and DeCarlo R (1995). Deterioration, autism and recovery in two siblings with follow-up. *Journal of the American Academy of Child and Adolescent Psychiatry*, 34, 232–37.

Pfefferbaum A, Sullivan EV, Swan GE, and Carmelli D (2000). Brain structure in men remains highly heritable in the seventh and eight decades of life. *Neurobiology of Aging*, 21, 63–74.

Piven, J, Berthier, ML, Starkstein, SE, Nehme, E, Pearlson, G, and Folstein, S (1990). Magnetic resonance imaging evidence for a defect of cerebral cortical development in autism. *American Journal of Psychiatry*, 147, 734–739.

Rakic P (1991). Plasticity of cortical development. In SE Brauth, WS Hall, and RJ Dooling, eds. *Plasticity of Development*. pp. 127–161, MIT Press, Cambridge, MA.

Rakic P, Bourgeois J-P, Eckenhoff ME, Zecevic N, and Goldman-Rakic PS (1986). Concurrent overproduction of synapses in diverse regions of the primate cerebral cortex. *Science*, 232, 232–35.

Rapin, I (1993). Autism: A Complex Developmental Disorder of Brain Function. Paper presented at the 25th Anniversary Celebration of The New York State Institute for Basic Research. Staten Island, NY.

Rilling JK and Insel TR (1999). Differential expansion of neural projection systems in primate brain evolution. *Neuroreport*, 10, 1453–59.

Rodier PM (2000). The early origins of autism. *Scientific American*, 282, 56–63.

Rumelhart DE and McClelland JL (1986). Parallel Distributed Processing: Explorations in the Microstructure of Cognition. The MIT Press, Cambridge, MA.

Schreibman L (1975). Effects of within-stimulus and extra-stimulus prompting on discrimination learning in autistic children. *Journal of Applied Behavior Analysis*, 8, 91–112.

Schreibman L, Charlop MH, and Koegel RL (1982). Teaching autistic children to use extra-stimulus prompts. *Journal of Experimental Child Psychology*, 33, 475–91.

Selfe L (1997). Nadia: A Case of Extraordinary Drawing Ability in an Autistic Child. Academic Press, London.

Smith T, Groen AD, and Wynn JW (2000). Randomized trial of intensive early intervention for children with pervasive developmental disorder. *American Journal on Mental Retardation*, 105, 269–85.

Stein DJ and Ludik J (1998). *Neural Networks and Psychopathology*. Cambridge University Press, Cambridge.

Stromland K, Nordin V, Miller M, Akerstrom B, and Gillberg C (1994). Autism in thalidomide embryopathy: a population study. *Developmental Medicine and Child Neurology*, 36, 351–56.

Tantam D, Monaghan L, Nicholson H, and Stirling J (1989). Autistic children's ability to interpret faces: a research note. *Journal of Child Psychology and Psychiatry*, 30, 623–30.

Thomas M and Karmiloff-Smith A (2002). Are developmental disorders like cases of adult brain damage? Implications from connectionist modeling. *Behavioral and Brain Sciences*, 25, 727–50.

Tirosh E and Canby J (1993). Autism with hyperlexia: a distinct syndrome? *American Journal on Mental Retardation*, 98, 84–92.

Werbos PJ (1974). Beyond regression: new tools for prediction and analysis in the behavioral sciences. Unpublished doctoral dissertation, Harvard University.

Xiang H, Lin C, Ma X, Zhang Z, Bower JM, Weng X *et al.* (2003). Involvement of the cerebellum in semantic discrimination: an fMRI study. *Human Brain Mapping*, 18, 208–14.

Zadeh LA (1965). Fuzzy sets. *Information and Control*, 8, 338–53.

Chapter 11

Conclusion: the future of neuroconstructivism

Mark H. Johnson and Denis Mareschal

In the course of our two volumes on neuroconstructivism we, and our contributing authors, have gone from outlining a general approach to cognitive and functional brain development, to detailing specific examples of how the general principles can be implemented. In Volume 1 a broad framework incorporating neuroscience, social and cognitive development, as well as computational approaches to neurocontructivism was provided. The last chapter of the previous volume reviews the framework and explores its implications and limitations in detail. In this second volume we have focused on computational modelling and robotic neuroconstructivist approaches, albeit that these efforts were often inspired and constrained by cognitive, behavioural and neuroscience data.

While the success of the neuroconstructivist approach will also depend on empirical techniques such as functional imaging, cognitive analyses and genetics, we have argued that computational modelling is of particular importance within the neuroconstructivist approach. Aside from the general importance of modelling discussed in both volumes, we maintain that it is a particularly important tool within the neuroconstructivist approach for several reasons. First, constructivist approaches to development have sometimes been regarded as somewhat mystical in comparison to the apparently clearer theories espoused by other schools of thought such as nativists. Demonstrations with formal modelling that complex representations can emerge without being prespecified help to dispel this myth. Such models can be analysed and principles extracted in a way that is generally superior to the verbal/logical/linguistic models used to champion nativist approaches. Another reason why modelling has a central role within the neuroconstructivist approach is that models can specifically bridge between traditional levels of observation (neural, brain systems, cognitive function; for more discussion of this point see Volume 1).

Finally, simulation work is vital for understanding the constraints imposed by body and environment on the emergence of representations.

Challenges

Computational modelling also faces a particular challenge when modelling data and theory within the neuroconstructivist approach. As discussed in Volume 1, and in many of the contributions in this volume, within the neurocontructivist approach understanding the effects of context (embrainment, embodiment, society) on the emergence of representations is critical. Contributors to this volume have illustrated the importance of context in a number of different ways: coupling between sensory and motor systems (Joanisse), modulation of some brain systems by other brain systems (embrainment, Sporns, Bednar), and interactions between brain, body and environment (embodiment, Schlesinger and Parisi, Demiris, Bednar). In some of the chapters in this volume, the body and external environment are simulated as well as the brain (Schlesinger and Parisi), while others believe that the 'real' constraints of a robot operating in a physical environment are essential (Demiris). One way forward for such modelling efforts is to increase the complexity of these models still further to assess multiple interacting contextual factors. However, while understanding the constraints imposed by context benefits from a modelling approach, this could potentially lead to models or robots that are so complex in themselves that they cease to be clarifying or explanatory. Particularly when simulating transformations that occur during development, there is no point in having a simulation that is as difficult to analyse as a real biological system.

We suggest that one way to overcome this problem is to limit the domain of modelling to that of detailed examples about which there is a wealth of neuroscience and/or cognitive and behavioural evidence. Indeed, this is an approach adopted by several of the contributors to this volume. One example comes from the 'ecological neural network' model of reaching designed to simulate the development of hand-eye coordination in young infants (Schlesinger and Parisi, this volume). In this simulation the authors suggest that no full internal model of the external world is required to attain successful reaching performance: rather, partial representations are sufficient. In these cases, combinations of contextual influence can be examined within a highly restricted domain of behaviour.

Principles

Another way to escape from the difficulty of knowing how much complexity to include in models is to build simulations designed to illuminate general underlying principles that can then be extended to other domains and levels (see also Rogers and McClelland, 2004, for a similar approach). Part of the aim of Volume 1 was to begin to generate a taxonomy of such explanatory principles. Contributors to this volume have provided some additional candidate principles/mechanisms that merit further consideration. For example, Raijmakers (this volume) proposes that 'bifurcations' (abrupt phase transitions) in brain, cognition, or behaviour are a hallmark of constructivist developing systems This poses a direct challenge to the implicit assumption that abrupt changes in cognitive or behavioural development must be due to the maturation of an underlying brain system, rather than the gradual accretion of knowledge through learning. Moreover, an important marker of bifurcations is the presence of hysteresis (van der Mass and Molenaar, 1992). That is, transitioning to a different (higher) level of performance affects the system as a whole such that the local transitions are not directly reversible.

Another example of a candidate underlying principle comes from Sporn's (this volume) principle of *'value'*. We interpret value as a special kind of embrainment that may have unique importance. Value refers to a modulatory brain signal that reflects the outcome or consequences of a behavioural interaction with the environment. While value initially occurs after a behavioural interaction, it may help brain systems predict the (positive or negative) outcome of future interactions, and thus in the long term contribute to proactivity dependence. Value also reminds us that embrainment is not necessarily confined to co-active regions of the brain, but that constraining influences on specialization can occur between regions whose activity is separated in time.

Micro-anatomy

As mentioned in the Introduction to this volume, there is a trade-off between simplifying reality and faithfulness to reality in computational models of the brain. Models that attempt to incorporate details of cellular structure and biophysics may not be appropriate for understanding at the neural systems or psychological level due to their greater complexity, and the difficulty of scaling such details up to large networks of neurons. On the other hand, abstract connectionist models ideal for cognitive analyses may miss crucial lower level constraints imposed by details of the neural architecture. These considerations raise the issue of how detailed the neural information incorporated into a model needs to be. The chapters in this volume vary from abstract cognitive models to more detailed implementations of neural circuits.

The neuroconstructivist approach to developmental disorders outlined in these volumes is commonly associated with a commitment investigating the consequences of 'micro-anatomical' differences such as different connection patterns, degrees of connectivity, or thresholds of activation of nodes (Joannisse, Cohen, and Chapter 11 of Volume 1). This commitment to investigating how widespread micro-anatomical differences can give rise to apparently specific patterns of cognitive deficit requires a level of modelling that implements at least some aspects of known neuroanatomy.

Environment

A related challenge for the future of modelling within the neurocontructivist framework is the extent to which the environment outside the organism needs to be real (as in robotics) or simulated (as in the Artificial Life perspective presented by Schlesinger and Parisi). In some ways, the structure of the external environment is as poorly understood as the constraints imposed by the body or other brain regions. However, we are encouraged that recent trends in sensory neuroscience are toward understanding the statistics latent in sensory input from the environment (Field, 1994). Fortunately, natural visual images tend to contain particular invariances (Sporns, this volume). In the future this work will need to be further extended from simple visual scenes, to the statistics associated with objects, categories of objects, social stimuli (see for example Mareschal *et al.*, 2000, or French *et al.*, 2004).

New directions

Building on the work presented in the two volumes of neurocontructivism, we see a number of future directions that could be fruitful to pursue. First, the effects of intrinsic and sensory-driven experience during prenatal life have, to our knowledge, only been addressed by Bednar and Miikkulainen (this volume). This approach potentially offers an insight into the origins of the propensities and biases evident in the newborn human infant. Second, the 'simulated genetics' approach taken by Schlesinger and Parisi is exciting, but given the importance of bottom-up constraints (refer Volume 1), we need to be cautious about whether the appropriate aspects of genetics, and gene expression, are being captured in these simulations and in their 'genetic algorithms'. Third, 'forward models' of motor control (Demiris, this volume) in which brain systems anticipate the outcome of actions before they have occurred have become influential in thinking about adult motor and cognitive performance. Such models move us more toward a view of the brain as an active agent in the world, and potentially offer insight into mechanisms of proactivity dependence during development.

There is also a need for more translational research that bridges between the basic-level questions that we have been describing here and applied intervention programmes or educational work. One of the great strengths of the chapter by Cohen is that his effort draws out explicit implications for effective interventions on children with autism from his models. Unfortunately, this is far too rare.

There is also a need to create developmental models that bridge between more than two levels of description. Most of the models that exist, including those in this volume, generally cross two levels of description only. That is, they use neural networks to show how the emergence of functional brain systems can produce the emergence of behaviours observed in children or infants. However, one of the main themes of Volume 1 is that we should strive for consistency across multiple levels. The single cell level should be consistent with the neural networks level which in turn should be consistent with the functional systems level and which itself should be consistent with the behavioural level. Most models only grapple with consistency across two of these levels. Although there are fascinating efforts to trace the developmental picture from the molecular to the behavioural level (e.g., Van Ooyen, 2003), each step along the route is examined independently. Ultimately, it would be helpful to see how three or four different levels could be superimposed within a single developing system.

A related issue is the need for developmental models that bridge between imaging data, neural functioning and behavioural data. Most models try to link the function of neural systems to behavioural development. Some models (especially in the adult neurophysiology literature) try to bridge between imaging data (such spike rates) and neural information processing. However, there are almost no models that try to bridge between all three areas within a developmental context. For example, while there are some models of the emergence of face processing abilities over the first six months of life (e.g., Acerra *et al.*, 2002), these are not related to the rich set of event-related potential data collected over the same age range. To do so would require developing accurate theories of how the measured imaging signals map onto the underlying neural information processing.

A final question is the extent to which ensocialment is taken seriously in any existing model. The robotics chapters in this volume certainly provide evidence of how embodiment is taken seriously, but there is very little work that we are aware of that looks at how social interactions, social scaffolding and explicit tutoring impact on development. There are some notable exceptions such as the work of Steels and colleagues examining vocabulary growth within a population of robots (Steels, 2003) or the developmental robotics efforts to give robots a realistic learning experience (Weng *et al.*, 2001). However, these remain relatively rare efforts that do not capture the rich fabric of social support that is available for the developing human child.

Final remarks . . .

In Volume 1 of *Neurocontructivism* we set out to investigate how the representations that underlie cognition emerge in the brain during development. We argued that the emergence of such representations is the outcome of a constructivist process involving constraints that operate at several levels of observation from the cellular environment to the social environment. Thus, to truly understand how these representations emerge we need to place brain, body and environment squarely at the centre of any explanation. In the final chapter of Volume 1 we also laid out what we saw as the most exciting implications, both theoretical and practical, of the neuroconstructivist approach. Among these were several lessons for how hands on research could be carried out within a neuroconstructivist framework. One such lesson was to incorporate more computational modelling within research programs exploring the origins of cognition. In this present volume we focused on examples of how computational and neural network modelling can be used as tools to explore factors interact to shape representations. It will be apparent to the reader that these are still early days for the neuroconstructivist approach to brain and cognitive development. We hope, however, that you leave this book as excited about its future as we are.

References

Acerra, Francesca, Burnod, Yves and de Schonen, Scania (2002) Modeling aspects of face processing in early infancy. *Developmental Science,* 5, 98–117.

Field, D (1994). What is the goal of sensory coding? *Neural Computation*, 6, 559–601.

French, RM, Mareschal, D, Mermillod, M and Quinn, P (2004) The role of bottom-up processing in perceptual categorization by 3- to 4-month old infants: Simulations and data. *Journal of Experimental Psychology: General*, 133, 382–397.

Mareschal, D, French, RM, Quinn, P (2000). A connectionist account of asymmetric category learning in infancy. *Developmental Psychology*, 36, 635–645.

Rogers, TT and McClelland, JL (2004) *Semantic Cognition*. Cambridge, MA: MIT Press.

Steels, L (2003) Evolving grounded communication for robots. *Trends in Cognitive Sciences*, 7, 308–312.

Van der Maas, HLJ and Molenaar, PCM (1992). Stagewise cognitive development: An application of catastrophe theory. *Psychological Review*, 99, 395–417.

Van Ooyen, A (2003*) Modeling neural development.* Cambridge, MA: MIT Press.

Weng, J, McClelland, J, Pentland, A *et al.* (2001). Autonomous mental development by robots and animals. *Science,* 291, 599–600.

Index